The Almond in California

R. H. Taylor

Alpha Editions

This edition published in 2024

ISBN : 9789366382173

Design and Setting By
Alpha Editions
www.alphaedis.com
Email - info@alphaedis.com

As per information held with us this book is in Public Domain.
This book is a reproduction of an important historical work. Alpha Editions uses the best technology to reproduce historical work in the same manner it was first published to preserve its original nature. Any marks or number seen are left intentionally to preserve its true form.

INTRODUCTION

The almond (*Prunus communis*) is supposed to be native to the countries around the Mediterranean and at present the bulk of the world's supply is produced in that region. It resembles the peach somewhat in manner of growth and character of blossoms and leaves, but the wood is much harder and the tree is longer-lived under equally favorable conditions. The fruit, instead of having a thick, fleshy pericarp as in the case of the peach, has a thin, leathery pericarp or hull, which splits on ripening and generally opens when dry, exposing the nut inside.

California produces over 98 per cent of the entire American crop and has done so for many years. During the period from 1900 to 1913 the number of bearing trees remained approximately the same, new plantings having replaced old orchards that were being pulled out. The variation in California production from year to year prior to 1915, as shown in figure 1, is due to seasonal variations rather than to change in acreage.

Imports into the United States from the Mediterranean countries are also shown in figure 1, the top line representing the total imports, the other lines, as indicated, showing the proportion of that total originating in the three principal countries exporting to the United States. Previous to 1912 the records of shelled and unshelled almonds were not kept separate. Since the records have been segregated, the percentages of shelled almonds imported each year have been approximately as follows:

	Per cent
1912-13	83
1913-14	70
1914-15	71
1915-16	82
1916-17	79
Average	77

With the 1915 crop the production in California entered upon what appears to be a long prospective increase. The large acreage of almonds set out in the last four or five years is the result of greatly improved market conditions due to the successful work of the California Almond Growers' Exchange. The first of these new plantings are now coming into bearing, and each year for many years in the future will continue to see increased yields. Large acreages are still being planted so that the almond production in California bids fair to continue to grow.

Within the next few years California growers will, in all probability, be forced to accept lower prices for their almonds than they are now receiving. The American markets are fully supplied at present prices, yet constantly increasing acreage will inevitably result in a greatly increased tonnage. European almonds are being produced at a lower net cost and can be laid down on the Atlantic Coast more cheaply than is possible with the California product. This brings the grower face to face with the necessity of becoming more thoroughly familiar with the most economical methods of production and marketing if they are to continue to make a profit. It is essential, therefore, that a careful study be made of all the factors concerned in the growth, production and final disposition of the almond crop.

HABITS

The almond is the first of the deciduous fruit trees to start growth and come into bloom in the spring, and normally the last one to shed its leaves in the fall. In other words, it has a very short period of rest. When the trees are forced into premature dormancy by mites or lack of moisture, they soon reach the end of their normal rest period before the winter season is over. Then the first warm weather in spring will bring the trees into blossom. In some cases where moisture and temperature conditions are favorable late in the fall, they may actually blossom before the winter season. In young trees that have become dormant unusually early, the rest period may terminate and then the tips of the branches resume growth and continue to slowly develop new leaves at the terminals throughout the winter. Trees which have been kept growing thriftily until the leaves have been forced to fall by the cold weather and frosts of winter, do not tend to blossom as early in the spring, nor do they open under the influence of a few days of warm weather in late winter or early spring.

Young trees blossom somewhat later than the older trees, and buds on sucker growth blossom later than the more mature portions of the same tree. The difference may amount to three or four days or almost a week. Well-grown trees carry large numbers of blossoms over the entire tree, as shown in figure 2.

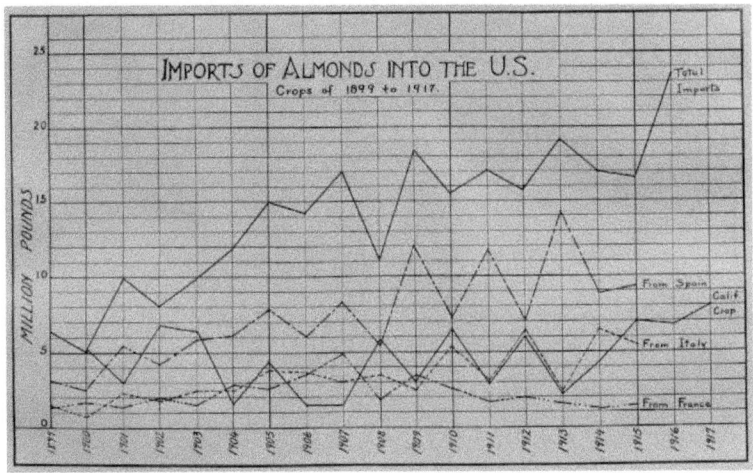

Fig. 1.—IMPORTS OF ALMONDS INTO THE U.S.
Crops of 1899 to 1917.

The wood of the almond is very hard and strong, enabling the tree to bear the weight of heavy crops where pruning has been given proper attention during the formative period of the young tree. As with other fruit trees, the almond is subject to heart-rot and care should always be exercised to prevent the checking and cracking of large wounds and consequent infection with decay organisms. The hardness of the wood makes it the finest kind of fuel, and when old orchards are being dug up the returns from the sale of wood often more than pay for the expense of digging and cutting up the trees and burning the brush.

The nuts are of two general classes—sweet and bitter almonds. The former is primarily the almond of commerce, though the latter is used largely in the manufacture of almond oil and almond flavoring, as well as in the manufacture of prussic acid. The bitter almond is also used largely in nurseries as a rootstock upon which to bud the almond and some other fruits.

For a long time there has been considerable evidence to show that some varieties are always self-sterile while a few are sometimes self-fertile. Work done in 1916 and 1917 by Tufts[1] shows that practically all varieties are self-sterile and that some of the self-sterile varieties are also inter-sterile. In these tests the principal commercial varieties were used. Blossoms of each variety were pollenized with pollen from its own blossoms and from each of the others. Checks were for natural pollination with each variety. The important results of this work are briefly summarized as follows:

The Nonpareil and I.X.L. are inter-sterile, although both are inter-fertile with the Ne Plus Ultra.

The Languedoc and Texas are inter-sterile.

The I.X.L. and Peerless are practically inter-sterile.

The California has proved the best pollenizer thus far tested, for all varieties that bloom near it.

The Drake is inter-fertile with the Nonpareil, I.X.L., Ne Plus Ultra, Peerless and Jordan, the only ones tested.

The I.X.L. is inter-fertile with the Drake, Jordan, California, Languedoc, Ne Plus Ultra and Texas.

The Ne Plus Ultra is inter-fertile with the California, Drake, I.X.L., Languedoc and Nonpareil.

REQUIREMENTS

While the almond is in many ways an easy tree to grow where conditions are favorable, it is more particular in its requirements than most common orchard fruits, and the grower may find it difficult to produce a good, thrifty tree unless he chooses the proper location. Very often it will grow well and make a fine healthy tree, but owing to unfavorable conditions, will not bear regularly, if it all.

Fig. 2.—Seven-year-old Ne Plus Ultra almond in full bloom, at the University Farm, March, 1915, showing distribution of blossoms well down into center of tree. Paper bags cover blossoms pollenized by hand.

CLIMATE

Heat.—Where the conditions of soil and moisture are favorable the almond will endure the intense heat of the interior valleys and even of the Imperial Valley, provided it is pruned properly to shade the main branches so as to prevent sunburn. Where trees, by severe pruning, are opened up suddenly to the intense heat of the summer sun, almonds will sunburn, but if the necessary opening up is done gradually, the bark will become inured to the new conditions without danger. The nuts grow and ripen more satisfactorily in the greater heat of the interior than along the coast.

Frost.—The almond tree is hardy and will endure fully as much cold as the hardiest peach without injury. Trees are found growing well in Illinois, Ohio, New York and other Eastern states. In very favorable seasons they may even bear fruit, though this happens very seldom, due to the extremely early habit of blooming before the spring frosts are over. The first warm weather seems to start the trees into bloom, especially where the enforced dormant season of winter is very long.

The blossoms, on the other hand, are very tender. There is a great range in the degree of frost which will cause injury, depending largely on the condition of the tree during the time that the fruit buds are forming and developing, as well as on the duration and severity of the frost. Buds and blossoms on trees which have been forced into premature dormancy, either by lack of moisture or by severe attacks of red spider, are much more susceptible to frost than those on trees which have continued growth late enough in the fall to provide for the proper development and maturity of the buds. After differentiation of fruit buds commences in the summer, the almond leaves should remain on the tree until late into the fall in order to strengthen and develop the fruit buds and store up the elaborated food material for the use of the buds in their normal development through the winter. Studies of almond buds gathered from healthy trees which held their leaves until late fall frosts at Davis, showed the first evidence of differentiation between fruit and flower buds commencing about August 18, while the flower was not completely developed until February 18 following. During the intervening time development proceeded unchecked through the winter even though the tree was apparently dormant. During the time the crop is ripening on the trees, little is done toward storing food material for the buds. If the leaves turn yellow or drop soon after harvest, the trees do not have the opportunity of storing a sufficient supply of plant food for their normal requirements and the buds are insufficiently nourished during the winter period. The resulting buds are weakened and the indications are that they are unable to endure unfavorable climatic conditions in the spring, such as light frosts, continued cold weather or sudden changes from warm to cold weather.

The most tender stage in the blossoming and development of the young fruit seems to be immediately following the dropping of the calyx lobes from the young fruit as it first commences to swell rapidly. The blossom becomes more and more tender as it opens out and reaches the above stage. After the young fruit has attained the size of a pea it rapidly becomes more resistant to low temperatures. Blossoms with the petals exposed but not yet opened have been known to stand temperatures of 24 degrees F. and blossoms with petals beginning to fall have stood 28 degrees F. No records are available as to the duration of these temperatures. In other cases, blossoms with the petals falling have been killed by temperatures of 30 and 31 degrees F. It must be remembered in this connection that the almond blooms earlier than other orchard fruits and, therefore, is often subjected to much more severe frosts than occur during the blooming period of the later fruits. The greatest injury is likely to occur when a frost follows one or more days of warm weather. When the mean temperature both day and night remains low, frosts that might otherwise kill the flowers or setting fruit do no harm. This is what occurred in February, 1917, at the University Farm, when repeated frosts at blooming time did no harm whatever.

In determining the desirability of a location in regard to its freedom from frost, the possibility of adequate air drainage is an important item. For this reason the lands along the lower foothills immediately above the floor of the valleys are ordinarily much less subject to frost—because the cold air is free to drain away to the lower levels. Generally the lands along the banks of streams which have been built up higher than the other lands of the Great Valleys through which they flow, are less subject to frost by reason of the natural flow of the cold air from them to the lower lands adjacent. For the same reason the planting of almonds in the lower lands of the valleys, no matter how large the valleys may be, should be avoided, unless the locality has been thoroughly tested for a long period of years and has proved to be an exception to the rule because of some peculiar situation with favoring air currents or air drainage, such as might exist near a natural draw in the hills where the settling of the cold air in some portions of the adjacent valley might be prevented. Such locations are generally confined to very small areas. Oftentimes an opening or draw in the hills may serve as an outlet for the drainage of much colder air from considerably higher elevations beyond, and then the danger from frost is very greatly increased. This is very common where cañons act as drains to conduct the cold air from the high Sierras to the valleys below.

Variable weather conditions, and especially as regards temperature in the spring after growth commences, are highly undesirable. Warm weather immediately followed by cold tends to produce sour-sap, fruit drop and kindred physiological ills. Oftentimes crops have been lost where no frosts occurred after blooming commenced, simply due to sudden changes in the weather. However, it is highly probable that the greater portion of the trouble with fruit dropping, when of the size of peas or larger, is due to improper pollination. When the soil is not sufficiently well drained at such a time, the sour-sap effect is greatly augmented.

Humidity.—Foggy or moist weather during ripening or harvesting is highly objectionable. The nuts do not dry out rapidly enough on the trees to prevent the growth of molds and consequent darkening of the shells. The nuts then require much heavier bleaching to brighten them properly for the demands of the market. The damp weather prevents the rapid and thorough drying-out of the kernel; the sulphur fumes are absorbed by the moist kernel and it sometimes becomes rancid before it is six months old.

Much damp weather in the spring encourages the growth of "shot-hole" fungus in the blossoms and fruit, often causing the loss of a considerable portion of the crop; the loss of leaf surface from the fungus infection is sometimes so great as to materially affect the vigor and vitality of the tree.

Rainfall.—It is impossible to state any definite amount of rainfall which will or will not maintain the trees and enable them to bear regular crops of nuts, for so much depends not only on the variation in rainfall in different sections and in different years, as regards distribution throughout the year, but also on the time and intensity of the fall, the character of the weather following the rains and the ability of the soil to receive and retain the rain that falls. Ordinarily, however, with the above factors favorable, it is conceded that where the winter rainfall averages sixteen inches, almonds can generally be grown without supplementing the water supply by irrigation, if the orchardist exercises reasonable care to conserve the moisture for the use of the trees. Where the rainfall is inadequate some means of irrigation must be found to make up the deficit.

In some sections the annual rainfall varies greatly from year to year. Often it falls in such a way that a large proportion of it is lost in the surface run-off. In many places the soil is so leachy that it is incapable of holding sufficient water for the use of the trees throughout the summer, much of the winter rainfall being lost in the underground drainage. Under either of these conditions, 40 inches of rainfall might not be sufficient. Very often winter rains are followed by desiccating winds so that a considerable portion of the rain which falls is lost by evaporation before anything can be done to hold it.

Continued rainy, damp and cold weather at the time of blooming is apt to sour the pollen or actually wash it away and thus prevent the fertilization of the blossoms, without which a crop is impossible. Bees and other insects are the principal means of accomplishing the pollination of almonds and such weather prevents them from working.

SOIL

The almond is a deep-rooting tree and draws heavily upon the plant-nourishing elements of the soil. In ripening the large number of seeds which it is required to do, the tree must draw upon a considerable area of soil in order to supply the large amount of mineral matter that is needed to develop and mature the seeds. Analyses of almonds, as compared with other commonly grown fruits and nuts, made by Colby,[2] show that the almond leads in the total quantity of mineral matters withdrawn from the soil. Colby further states that "The stone fruits fall much below the almond in total ash (mineral matter) excepting the olive, the ash of which, however, is largely silica (nearly eight-tenths), an ingredient so plentifully distributed in all soils that it is of no pecuniary value." Table I illustrates this statement.

These figures suggest the necessity of having a deep, rich, well-drained soil for best results. For this reason and because of the deep-rooting habit of the almond, the soil should be at least ten or twelve feet deep.

Hardpan.—Compacted substrata in the soil, whether they be hard clay layers or cemented layers of silicious, ferruginous or calcareous origin, are objectionable. They not only prevent the roots from foraging to a considerable depth as they normally tend to do, but they prevent proper drainage and aeration of the soil. If such layers are comparatively thin, that is, not more that two or three feet thick at the most, they may be shattered with dynamite so as to allow the moisture, air and roots to penetrate to the better soil below. Hardpan, therefore, should be avoided where it is too thick to be broken up or where it is not underlaid by desirable soil.

TABLE I
SOIL INGREDIENTS EXTRACTED BY THE ALMOND AS COMPARED WITH OTHER ORCHARD TREES, AS SHOWN BY ANALYSIS OF 1000 LBS. EACH OF THE CROPS IN A FRESH CONDITION

Fruit	Potash, lbs.	Lime, lbs.	Phosphoric Acid, lbs.	Total Ash, lbs.	Nitrogen, lbs.
Almond (hulled)	5.49	1.72	4.33	15.00	16.40
Almond (not hulled)	9.95	1.04	2.04	17.29	17.01
Walnut (hulled)	1.50	1.81	2.78	7.50	10.20
Walnut (not hulled)	8.18	1.55	1.47	12.98	5.41
Chestnut (hulled)	3.72	.71	1.89	8.20	8.00
Chestnut (not hulled)	3.67	1.20	1.58	9.52	6.40
Prunes (green)	2.66	.13	.53	4.03	1.48
Apricots (green)	2.83	.18	.71	5.16	2.29
Olives	8.85	2.32	1.18	94.63*	5.85

* 80.7 pounds of which is silica.

Humus.—A plentiful supply of humus in the soil is essential. It not only improves the physical condition of the soil, but assists drainage, moisture retention and in rendering the plant food available in sufficient quantities for the use of the trees and for the maturing of full crops of almonds. Many orchards have been very light producers year after year because of a deficiency of humus in the soil.

Drainage.—The almond root is very particular as to its air and moisture requirements in the soil. It will not endure standing water in the soil for any length of time, especially during the growing season. Exclusion of air by excessive moisture is believed to be one of the most productive causes of "sour-sap." If allowed to continue for any length of time such conditions will cause the death of many or even all the roots and with them the top.

Water Table.—A factor which is very commonly overlooked in connection with the natural drainage of almond lands is the position of the water table at different seasons of the year. Great care must be exercised in choosing a location to be sure that the water table does not rise during the summer. This is a very serious problem in many irrigated sections. Where the water table during the winter months is less than 12 feet in depth it is highly desirable to have as little fluctuation as possible. Where fluctuations take place at a greater depth than 12 feet they are not generally serious. The ideal condition is where the water table is highest in the winter and quickly drops after the winter rains are over, to a depth of from 10 to 12 or 15 feet, remaining at that point during the remainder of the growing season.

The soil in addition to being well drained, must be sufficiently retentive of moisture to supply the tree throughout a long, dry growing-season. If the soil will not retain a sufficient amount of the winter and spring rains, recourse must be had to irrigation to supply the deficiency.

Alkali.—Alkali lands are unsuited to almond culture and should be carefully avoided.

In summarizing the soil requirements for almond culture, it may be stated that the ideal almond soil is a medium loam, uniform in texture, or nearly so, to a depth of at least twelve feet, well drained and yet retentive of moisture for the use of the tree during the summer. Fortunately some of the best almond soils are situated along stream banks where the land is relatively high, and is, therefore, less subject to frost. These streams flowing from the mountains and foothills have built up their own beds by the detritus brought from the hills. The coarser particles being deposited first and nearest the stream itself, make the better drained soils, while the finer particles and the clays, deposited further back from the bank and in the lower lands, form the heavier soils.

The various conditions mentioned above are what the tree should have for best conditions of growth and production. Oftentimes these conditions may be approached without the soil being as deep as ten or twelve feet. Exceptions to this will be mentioned in discussing the various almond districts of the state. It is essential to understand that trees, while growing and bearing on shallow soils in some localities, do so because of other exceptionally favorable conditions; either the soil is exceptionally well drained and yet sufficiently retentive of moisture, or the humus in the soil is plentiful and the roots are able to work into the underlying partially decomposed rock for moisture and some plant food. In such localities the trees bear comparatively well because of the exceptional freedom from frost in the spring. Trees in these localities are generally smaller than on the deeper, richer soils, and where other conditions are equal, they bear crops in proportion to their size.

ALMOND DISTRICTS

Almonds are grown in nearly every county in California. In some counties the few trees growing only occasionally succeed in producing a crop of nuts. There are sections in nearly all parts of the state, however, where they are a success commercially. Within these sections may be found desirable and undesirable locations, depending upon soil and moisture conditions and freedom from injurious frosts. Any discussion of a district, therefore, does not necessarily mean that all lands within that district are uniformly adapted to almond culture. On this account it is impossible to define a district any more closely than to name the center and include with it the outlying districts. In the same way it is sometimes impossible to say just where one district begins and another ends. Adaptability of any special location can be determined only by careful study of the land itself and diligent inquiry of those familiar with it.

As far as possible, districts should be chosen where a definite cold winter season exists. Warm weather and lack of freezing temperatures do not hold the trees fully dormant and any frequent occurrence or unusual continuation of spring weather in the winter will start the trees into growth; cooler weather following, interferes with the normal flow of sap, results in injury to the tree and blossoms, and often causes gumming of the nuts which mature. This condition exists largely in the lower elevations in Southern California and especially in the coastal portion, where the ameliorating influence of the Pacific is felt. This same condition exists close to the coast in the northern portion of the state. Further inland and at higher elevations the winters are more pronounced, and where these are not too severe or prolonged the almond thrives best.

SACRAMENTO VALLEY AND FOOTHILL SECTIONS

Commencing in Solano County, about eight miles northwest of Suisun and extending as far north as the southern end of Shasta County, close to the base of the foothills on the west side of the Sacramento River, are a large number of orchards on the bottom lands of the valley. These are found principally on the higher lands along the banks of streams flowing from the Coast Range in a generally easterly direction to the Sacramento River. The two most important streams of this type are Putah Creek, forming the boundary between Yolo and Solano counties, and Cache Creek, flowing through the Capay Valley and across Yolo County to the Sacramento River.

These plantings being on the floor of the valley are more subject to frost than the foothill plantings, but they have the advantage of being on the rich valley soils, and while they may lose a crop occasionally from frosts, they make it up in the long run by the much larger growth of the trees and their ability to produce larger yields. Many orchards do not show any such ability to produce large crops, but such a failure must be attributed to lack of pruning and care, rather than to any inherent inability of the large trees to produce nuts. Again, trees on the rich and moist bottom lands may be kept so vigorous and healthy by good care that they are apparently able to endure lower temperatures than less vigorous trees on the shallow and poor upland soils where frosts are less common or severe.

Many recent plantings have been made along the lower foothills on the west side of the Sacramento Valley. The three most important districts where these plantings have been made are west of Dunnigan, Arbuckle and Corning. The plantings in these districts are mostly on a gravelly clay or clay loam soil which is not as rich nor generally as deep as the bottom lands and, consequently, the trees are somewhat smaller, but they have the advantage of freedom from spring frosts due to superior air drainage. The problem in this district is chiefly one of moisture retention by cultivation, as in most cases the growers are unable to get water for irrigation at a reasonable cost, if at all.

The valley districts extend to Tehama County and down the east side of the Sacramento River, the same as on the west side, the principal centers being at Chico, Durham, Liveoak, Pennington and Sutter. Parts of the Liveoak section are too low and many orchards have suffered from poor drainage both of water and air.

The foothill sections on the east side are principally around Antelope, Fairoaks and Orangevale in Sacramento County and the southern end of Placer County. Here the danger from frost is slightly greater than on the western foothills because of the proximity to the snow-covered Sierra Nevada Mountains. On the other hand, water from the Sierras renders irrigation possible at a reasonable cost, so the trees can be kept in good condition. Care must be exercised here to avoid frost pockets. There is also danger of poor soil drainage in the swales.

SAN JOAQUIN VALLEY AND FOOTHILL SECTIONS

Conditions here blend very much into those of the Sacramento Valley sections. Continuing south there are plantings around Acampo, Lodi, Stockton, Linden, Ripon, Modesto and more scattered plantings farther up the valley. The danger from spring frosts increases southward due to the earlier blooming of the trees. In some of these sections, notably around Ripon and Modesto, the excessive use of irrigation water for alfalfa and other crops has resulted in a rise of the water table in many places until it is within a foot or two of the surface during the summer months, and in many other places comes to within four feet of the surface.

On the west side of the San Joaquin Valley the principal plantings have been in eastern Contra Costa County. Most of these orchards are of old trees, planted from 15 to 30 years ago. In order to obtain satisfactory air and soil drainage the orchards were planted on the rolling hills, the soil being nearly all blow-sand. Irrigation is too expensive to be installed on most of this land, and the problem in that section is to conserve moisture and at the same time hold the soil in place and prevent it from being carried away by the wind.

Fig. 3.—Typical hillside orchard of Jordan almonds near Los Gatos. Trees are variable in size, and some are missing. In the right foreground is a typical spot of missing trees resulting from Oak fungus infection.

THE COAST SECTION

Almonds were planted quite extensively in past years in many of the coast valleys, such as the Santa Clara, and where they were protected from the direct influence of the moist ocean breezes, they apparently did well. Later most of these orchards were replaced by prunes and apricots as they were generally found to be better paying crops in these valleys. On the hills, however, many typical hillside orchards remain, as shown in figure 3.

Localities directly affected by the coast breezes have proved themselves to be unfit for commercial almond culture.

INTERIOR COAST VALLEYS AND HILLS

Back from the coast in the smaller valleys and on the hills where the coast fogs seldom reach, many locations may be found where almonds are producing successfully. These favorable localities are scattered, owing to lack of proper soil or moisture conditions.

Paso Robles District.—The one outstanding district in the coast hills where the almond is being grown successfully is in the country west of Paso Robles. There, at an elevation ranging from 900 to 2000 feet, where there is sufficient air drainage to the considerably lower land near by, the almonds are doing well. These plantings are, in most cases, on a comparatively heavy soil, sometimes approaching adobe, and underlaid by marl. This limestone formation is quite permeable to both water and the roots of the trees, and the rock itself acts as a sponge and holds considerable quantities of moisture. On the other hand, consisting of steep hills, the land is excellently drained and the trees seldom suffer from standing water around the roots.

The chief objection is that most of this land is very shallow, generally ranging from two to four or five feet deep. In some places almonds are found growing where the marl is so close to the surface as to be turned up by the plow. This makes the retention of moisture for the use of the trees a serious problem. The trees do not attain large size on such soils and the nuts are inclined to be somewhat smaller than normal. The steep character of much of the land makes tillage and spraying rather difficult and expensive. As long as the price of the land is kept down to a reasonable figure, the grower can afford to put more money into the care of both land and trees. It must be thoroughly understood that there are many localities close to Paso Robles that are greatly subject to frost and hence not at all suited to almond culture. This is especially true of the lands directly around and east of Paso Robles, and also of any of the comparatively low lands throughout the district.

SOUTHERN CALIFORNIA

The entire coast district of Southern California has practically proved itself to be unfitted for almond culture, except at elevations above 1500 or 2000 feet. The limiting factor seems to be the lack of a sufficiently definite winter season at the lower elevations. Trees are inclined to bloom too early in the season or before the winter is over.

Above 1500 or 2000 feet the increased elevation gives a sufficiently definite cold winter season and the trees tend to blossom at more seasonable times. Aside from scattered plantings in the hills of San Diego County and a few orchards around Hemet and San Jacinto in Riverside County, the principal plantings are at Banning and in the Mojave Desert.

Banning District.—Within a radius of three miles of Banning, with an average elevation of 2300 feet, where a plentiful supply of water is to be had from the mountains to the north, are a large number of orchards ranging in age from 2 to 29 years. The bulk of the plantings were made during 1910 and 1911. This is the oldest district in southern California and has proved itself to be well fitted for almond culture.

Antelope Valley and Vicinity.—The old plantings in the Mojave Desert are few and far between. They have, however, served as an impetus for large plantings in recent years many of which have turned out to be failures. This happened in many portions of Antelope Valley, west of Lancaster, on the Southern Pacific Railroad in northern Los Angeles County. Many more of these plantings are young and still have to prove themselves.

Orchard almond trees are never grown from seed, as they do not reproduce true to type, but are propagated by budding desirable varieties on seedling roots in the manner commonly employed in nursery practices for other stone fruits, like the peach. Every grower must decide, however, on what rootstocks he will plant his trees.

ROOTSTOCKS

There is much to learn yet regarding the behavior of the different roots under varying conditions when used as stock for the almond, but sufficient information has already been gathered to permit of certain recommendations being made.

Almond Root.—The greatest proportion of almonds in this state are on almond roots. Where the most desirable soil, moisture and drainage conditions exist it is best to select the almond root. It will live in comparatively dry soils, but cannot be expected to make a tree of large size or bear crops of nuts if the dry conditions continue. Where irrigation is not available, and the soil is deep and of proper texture, sufficient moisture may be retained in most years by cultivation to enable the tree to grow and bear fruit. When extra dry years come, the almond root may not enable the tree to grow or produce any better than would another kind of root, but it will carry the tree over a few dry seasons as well or better than any other. Then, when sufficient moisture does come, the almond root is ready to start the tree off in its normal course with the least loss of time.

Fig. 4.—Languedoc almond on peach[3] root; typical of other trees in same orchard, near Davis. Planted about forty-seven years before. Compare with trees in same orchard on Myrobalan root, as shown in Fig. 5.

The bitter almond is often recommended as being superior to the sweet almond as a stock. This claim has never been proved. Experiments indicate that there is fully as much variation between the bitter almond seedlings grown from seed from different trees as there is between seedlings from bitter and sweet almonds. Both are equally subject to attack by gophers. The grower's chief object, therefore, should be to secure as healthy, vigorous trees as possible whether they be on sweet or bitter almond roots.

Peach Root.—Where the soil moisture is quite variable in different portions of the soil, or variable from time to time during the growing season, the peach root will probably be most satisfactory. Soils less than six or eight feet deep, or where gravelly or hardpan layers occur at less depths, are not satisfactory in most cases for the almond root, and under these conditions the peach root is better. The peach root is a little better where irrigation is practiced during the summer, and especially so where there is danger of slow drainage of excess water from the soil. The peach does not thrive with standing water around its roots, but will better withstand fluctuating or temporary extremes in water supply than the almond. Its union with the almond is entirely satisfactory. See figure 4.

Fig. 5.—Languedoc almond on Myrobalan root, typical of many other trees in same orchard, near Davis, planted about forty-seven years ago. Compare with trees planted same time on peach root as shown in Fig. 4.

Davidiana Root.—Within the last five or six years the United States Department of Agriculture has introduced the seed of a Chinese peach, *Prunus davidiana*, the root of which is more resistant to alkali than the ordinary peach. This has been found to unite readily with the almond, although experiments have not yet gone far enough to fully determine its true value. It gives promise, however, of being a very desirable stock for sections where alkali may be troublesome.

Undesirable Roots.—The Myrobalan plum has frequently been recommended as a stock for the almond on poorly drained soils. The two make a strong union, but the plum root grows much slower than the almond top, as indicated in figure 5. A comparison with figure 4, in which both trees are 48-

year-old Languedoc, shows that the trees on Myrobalan root are pinched-in below the union while those on peach root are swollen at the base. Even young trees show the same pinching-in below the union. Almond trees on Myrobalan root do not make as large trees nor do they bear satisfactory crops in any portion of the state where it has been possible to compare them with the same varieties on other roots in similar situations.

The apricot is occasionally recommended because of its strong, thrifty growth, but it is a mistake to attempt to use it, as the union is not satisfactory. The trees grow thriftily for awhile, but before they reach full bearing they either break off at the union or are blown over above the union by moderate winds.

TOP-WORKING OLD TREES

Often through injudicious selection of varieties for original plantings, or for some other reason, it becomes desirable to work over old trees to more desirable varieties. This may be done by budding or grafting. In either case the tree may be entirely reworked in one season or one-half may be done the first season and the other half the next. The almond will recover fully if completely deheaded to a height of from five to six feet from the ground, provided it is not subjected to severe winds or excessive moisture in the soil. Winds are liable to break off the new top during the first year or two. Where these are dangerous it may be well to leave half of the tree on the windward side to be worked the second season. The remaining portions will help to serve as a wind-break until the top-worked portion of the tree has a chance to unite solidly with the stub of the main branch to which it is attached. Where winds are not likely to do damage it is generally the most economical method to work over the entire tree at one time. If the roots are subjected to excessive moisture conditions during the first season, the new growth may be unusually vigorous and much more likely to be blown out as a result of the heavy top and the poor attachment to the stub. "Sour-sap" may also be very serious. Under such conditions leaving half the tree the first season will enable it to distribute the growth more evenly with less undesirable forcing.

Whether the top-working shall be done by budding or grafting depends largely upon the wishes of the grower and the skill of the workman. Grafting will give a new tree more quickly than budding, because by the latter method a year is lost in growing a supply of new wood on the stubs into which the buds must be placed. By grafting, the new scions may start growing the first spring without any delay. Ordinarily the best method is to graft the trees the first winter and then, where the grafts fail to grow, buds may be inserted in the new growth which will come from the stub. During the first two seasons especially, great care must be exercised to keep out the sprouts that interfere with the growth of the scions. At the same time, all water-sprouts should not be removed during the first summer or there will not be sufficient leaf surface to perform the necessary functions of the tree. In addition, the scions tend to grow so rankly that they will be top-heavy and much more liable to be blown out by the wind, as shown in figure 6. This may largely be prevented by pinching the ends of the new scion growth during the early summer to force lateral branching. At the end of the first year all water-sprouts should be removed leaving nothing but the growth from the scions.

Fig. 6.—Twenty-seven-year-old trees deheaded two years before, showing the break-off of the new growth by strong winds.

PLANTING

The usual preparation given land for orchard purposes prior to planting should be applied to land being prepared for almonds. Special care must be given to insure thorough aeration of the subsoil by breaking up all hardpan, plow-pan or other compacted layers in the soil, where possible.

Fig. 7.—Almonds planted twelve feet apart. Trees in very weak condition and almost entirely defoliated by mites before harvest. The few nuts ripening on the trees are small "sticktights." Photo taken September 21, 1915.

Distance.—Almonds in most soils should be planted 30 × 30 feet on the square. In rich, deep soils the trees quickly fill the intervening space, the roots occupying the entire soil area long before the tops touch. In poor, shallow soils, or soils deficient in available moisture, the trees may not occupy the entire area above ground, but the roots will require more horizontal feeding space to enable them to make the size trees and bear the crops that they should.

Trees planted too close together often abstract the moisture from the soil before the growing season is completed. This shortage of moisture, with the aid of mites, commonly called red-spider, causes the loss of the leaves long before they normally should drop. Under such conditions the fruit buds are unable to make the vigorous growth which they should. These weak buds are unable to endure the degree of frost or other unfavorable conditions that stronger ones withstand without serious injury.

The trees must have plenty of sunlight and air. If planted too close, the trees tend to grow too high, each one striving for the light which is only available from above. In figure 7 the upward tendency of the trees is clearly shown. Such trees, if pruned as they ought to be, have a much greater tendency to send out numerous water-sprouts than trees which have plenty of room for the tops to expand laterally. After the tops of the trees interlock and shut out the sunlight from the lower portions, the smaller branches and fruit spurs in those parts gradually weaken and die, and eventually the entire crop is produced on or near the tops of the trees, where direct sunlight is available, as shown in figure 7. The excessive upward growth of the trees, with the consequent forcing of the fruit bearing to the top, not only greatly increases the difficulty and cost of pruning, spraying and harvesting, but reduces the possible bearing surface of the trees.

Setting the Trees.—The utmost care is necessary in setting out the trees to secure a uniform stand of vigorous trees. The trees are planted much the same as other deciduous orchard trees, but care must be used to spread the roots well to secure a thorough compacting of the soil around all the roots, and to see that the trees are not planted deeper than they were in the nursery.

CULTURE

Soil Handling.—The almond is one of the most exacting of fruits as regards its cultivation. The assumption is very often made by growers that because the tree will live through periods of prolonged drouth, it will also thrive under careless or poor methods of cultivation. That this is entirely erroneous may be inferred from the fact previously referred to that the almond draws more heavily upon the plant food of the soil for the ripening of its crop than any of the common orchard fruits. Experience has abundantly proved that many almond orchards are not bearing profitable crops because the requisite soil constituents are not available in sufficient quantities during the long growing season. The reasons for this condition may be one or several. First, there may not be sufficient moisture available throughout the season, or it may not be distributed evenly or in sufficient amounts throughout the entire soil area. Second, there may be insufficient aeration of the soil; and third, there may not be sufficient humus to fix the soluble plant food in the soil and render it readily available as needed by the roots.

Proper distribution of moisture in sufficient quantities is essential to secure adequate solution of the mineral elements needed by the roots for the use of the tree in all its parts. Sufficient aeration is needed to provide for the normal oxidation and weathering of the soil particles, and also for the respiration of the roots themselves.

There seems to be no question about the value of spring plowing and cultivation to put the soil in good physical condition, but many growers apparently question the value of summer cultivation. Though the surface of the soil appears to be unchanged after a period of several months, the fact is entirely overlooked that the structure of the mulch has been gradually changed and capillarity to the surface has been restored. The result is that evaporation takes place so rapidly from soils in such a condition, as well as from the leaves of the trees, that long before the end of the season the moisture is practically gone.

The methods and tools used are much the same as for other orchard trees and it is only necessary to call attention to items likely to be neglected. Spring plowing should vary in depth from year to year to avoid the formation of an impervious plow-sole. The depth to plow ranges from a minimum of six inches to ten or twelve inches or more as desirable. Summer cultivation should be kept up at least once a month, and preferably oftener, throughout the summer months and the soil should be stirred to a depth of four inches to provide a mulch sufficient to hold the moisture effectively in the hot, dry climates where almonds are grown. The actual number of cultivations necessary will depend on the types of soil. Harness with projecting hames or broad singletrees or with projections of any kind to catch on the branches or bark of the trees should never be used in an orchard.

Cover Crops.—Constant cultivation throughout the summer allows the humus to be burned out of the surface soil and, by hindering the growth of vegetation, prevents the addition of a natural supply of humus to replace that which has been lost. It is necessary, therefore, that some artificial means be used to supply the deficiency. The annual growth of winter cover crops to be plowed in during the spring, while not entirely replacing the humus burned out in the summer, serves to reduce the annual loss and at the same time assists materially in improving the texture of the soil.

A shade crop, such as alfalfa, may be grown successfully in mature orchards where irrigation water is available, and where perfect drainage of surface and subsoil can be maintained. During the first four or five years or more such a crop is generally a detriment to the almond trees, but after the trees reach maturity it may be beneficial in securing better drainage and aeration of the soil by the penetration of the numerous roots to considerable depths and their consequent decay when it is plowed in. Alfalfa also supplies nitrogen to the soil and thus gradually increases the fertility. The great danger with such a crop is that the orchardist is apt to irrigate for the alfalfa at the expense of the trees because of the quicker returns from the former. In deep, rich soils the cuttings of alfalfa may be removed from the land, though the last cutting in the fall should be allowed to remain on the land. Where this is done, a natural mulch is obtained which helps to increase the supply of humus. If the soil is poor and lacking in humus it is better to leave more cuttings of the alfalfa on the ground to decay. As this continues the soil becomes, through the increased accumulations of humus, more permeable and yet more retentive of moisture, and less irrigation will serve the trees more effectively.

During the early spring a well-established stand of alfalfa may assist so materially in getting rid of the excess water in the soil that sour-sap, fruit drop and kindred ills resulting from the unseasonable warm weather while the ground is filled with water and the trees are starting into growth, may be greatly reduced or avoided entirely.

Irrigation.—Water for irrigating almonds must always be used with great care whether it be in summer or winter. All water should be so applied that it will readily spread through the soil and not remain in excessive quantities for any length of time. Water should not be applied during blossoming or setting of the fruit in the spring or within a week or more of harvest. On the other hand, water applied just before the hulls commence to open, where the soil has become too dry, greatly reduces the number of sticktights, or almonds in which the hulls stick to the ripening nuts without opening.

Fertilization.—Little or no systematic work has been done up to the present time in the use of commercial fertilizers on almonds. The use of barnyard manure is desirable wherever obtainable. The value of almond hulls as fertilizers is doubtful because of the difficulty of bringing about their thorough decay when applied in sufficient quantities to be worth while. The use of lime to correct soil acidity and for improving the texture of heavy soils will be found desirable where this is obtainable at reasonable rates.

PRUNING

The almond trees normally set a larger number of fruit buds than they are capable of maturing. The number of blossoms set on a healthy tree to produce a full crop will be generally not more than 20 per cent of the total number which opened in the spring, and oftentimes much less. It has frequently been noted that trees which are somewhat lacking in vigor are inclined to set a larger number of fruit buds than those that are strong and vigorous. The normal set of blossoms on a young, vigorous tree is shown in figure 2. The result is that with the sub-normal strength of such trees being distributed among the extra number of fruit buds, the latter are not able to develop into the strong, healthy buds they should. The results of such conditions have been discussed on pages 23, bottom, and 24, top. One of the best means of securing and keeping healthy, productive trees is by careful, thoughtful and systematic pruning. The treatment given most of the almond orchards would indicate that the growers consider pruning to be one of the least important of the cultural practices.

Fig. 8.—Typical Drake almond tree in University Farm orchard; nine years old.

Fig. 9.—Typical Languedoc almond tree in University Farm orchard; nine years old.

To be most effective, pruning must be followed systematically throughout the entire life of the tree. The details of the practice will vary greatly, depending on the soil, moisture and climatic conditions surrounding the tree, as well as other treatment accorded it. At planting time, it is generally agreed that the tree should be cut back to between 18 and 24 inches from the ground to reduce the top to correspond to the shortened root system and to form a low head for the tree. A low head is desirable to avoid sunburn of the trunk and to keep the tree within easy reach for pruning, spraying and harvesting. During the first summer in the orchard much can be done to obtain a well-balanced head, to get the main branches well spaced on the trunk, and to prevent crowding later on, by pinching out all undesirable growth as fast as it appears and forcing the growth into those parts of the tree which are to remain. If these branches tend to grow too long and willowy, they may be made to branch by pinching back when they attain the desired height. In this way, if the tree can be kept growing vigorously, the beginning of an excellent foundation framework may be developed the first year.

Fig. 10.—Typical Nonpareil almond tree in University Farm orchard; nine years old. Note numerous water sprouts which must be removed.

Fig. 11.—Typical I.X.L. almond tree in University Farm orchard; nine years old.

The following winter, when the tree is one year old in the orchard, if it has grown too vigorously and the branches are too long and willowy and not properly branched it should be headed back to improve the shape and symmetry of the whole tree. If the tree has not made sufficient growth to give it a properly shaped head, it should also be cut back even more severely than in the case of the too vigorous growth. There sometimes will be trees that during the first year will develop such a well-shaped and stocky frame that no further heading will be necessary, all future pruning being limited to thinning out undesirable and interfering branches. Ordinarily, however, it will be highly desirable or necessary to continue the practice of heading-in the yearling tree at least. As long as heading continues it will pay to give special attention to pinching and thinning the following summer. In this way the frame of the tree may be developed more quickly and the future necessity of heading be materially lessened. In rich bottom lands where moisture is plentiful, heading may be necessary when the tree is two, three and, in extreme cases, 4 years old, in order to spread the tree and prevent it from growing too high. The necessity for this can only be determined by a careful study of the tree itself and the conditions surrounding it.

Fig. 12.—Typical Ne Plus Ultra almond tree in University Farm orchard; nine years old.

The habit of growth of a given variety will have a great deal to do with the number of seasons during which heading back will be practiced. Spreading varieties, such as the Drake (see figure 8), if making a strong, vigorous growth, should not be headed any longer than is necessary to start a sufficient number of main branches. Every opportunity must be given the trees of this variety to grow as upright as possible. Summer pruning of the drooping, undesirable branches wherever they develop, will help to increase the growth of the upright branches. By pruning as far as possible to inside buds, every opportunity will be given the trees to grow upward. If trees of such varieties make a poor, weak growth, a very heavy heading will force them to make a stronger and more vigorous growth the following season, provided any unfavorable soil and moisture conditions are corrected. Care must be exercised with these trees to prevent the downward growth of any of the branches by pruning them off during the summer, as suggested above. As soon as a strong, upright growth is started, no further heading should be done, but careful thinning by removal of undesirable growth should be continued.

Upright varieties, such as the Languedoc (see figure 9) and Texas, may require heading for a longer time than the more spreading varieties. It is necessary to force them to branch nearer the ground and they should be made to spread as much as possible. Pruning to outside buds and summer pinching will aid materially and make it unnecessary to head back after the first two or three years. Where this is not done, heading may be necessary at the end of the third and possibly to the fourth year if the tree is in rich, deep, moist soil. Another advantage of summer pinching and removal of water sprouts is that the trees are not allowed to develop a thick "crow's nest" at the end of the stubs where heading back was done the previous winter, making the following winter pruning much easier. The habit of growth of each of the principal commercial varieties is shown in figures 8 to 12, inclusive.

After the main framework is properly started, further pruning will be limited to thinning out the tree to keep it sufficiently open and to remove all dead, injured and interfering branches. Most of the growth after this will continue from the ends of the remaining branches and as the trees grow higher they will tend to branch more. All cross branches and those that are liable to interfere later should be taken out. All water sprouts should be removed unless they are needed to fill in undesirable open spaces.

As the trees come into full bearing and approach maturity, the vigorous wood growth will cease and will be replaced by a moderate yearly growth of wood which will gradually increase the size of the tree without making it unmanageable. Where trees continue to make excessive wood growth when they should be bearing, the trouble may generally be traced to unusually rich, moist soil, a condition which pruning cannot correct. It may be beneficial to stop pruning altogether for a few years, which will be likely to throw the trees into heavy bearing and thus stop excessive wood growth. After the bearing habit is once thoroughly established, moderate pruning may be done to get the tree back into proper shape.

Fig. 13.—Eighteen-year-old almond trees deheaded six feet from ground, in the Armstrong orchard near Davis. Stubs painted with white lead. Photo taken February, 1914. Compare with Figs. 14 to 19.

Fig. 14.—Same trees as Fig. 13, showing growth one year after deheading. Photo taken March, 1915. On some of the trees long stubs were left near the bottom of the trees from which new growth never started. Only a moderate thinning out of the new growth was done.

The mature tree should have its branches so spaced that sunlight may penetrate to all portions of the tree. This is necessary for the continued health and life of the fruit spurs in the lower portions of the tree. If the top is too dense, the new growth can only continue in the top and outer portions of the tree where sunlight can penetrate. When an adventitious bud does start in the center of the tree it immediately makes a strong vigorous growth to reach the sunlight far above. Such a growth is called a water sprout. If the trees are kept so pruned that the shadow on the ground during the greater part of the day is mottled with sunlight while the trees are in full leaf, the small growth may be maintained in the center and the bearing surface greatly increased. Where this is done the trees will not tend to grow as high as they otherwise would. Pruning of healthy, mature trees, therefore, will be limited to thinning out and removal of dead and interfering branches and water sprouts. Portions which grow too high may be brought down by removal close to a shorter and lower branch. By giving this treatment only to small portions of the tree during a season, the forcing of water sprouts may be largely avoided.

Fig. 15.—Same trees as Fig. 14, showing two seasons' growth after deheading. Photo taken February, 1916, just after pruning had been completed.

Old trees, or those which, through improper care or neglect have become weak and unfruitful, may often be rejuvenated. One method is to dehead them to a height of about six feet from the ground and to grow entirely new tops. By this method about three years are required to bring the trees back to the point where they are again ready to bear profitable crops. The objections to this are that it is an expensive operation, no crops whatever can be obtained for two years, organisms of decay may gain entrance to the wood, and sometimes the sudden shock to the tree renders it more subject to physiological disturbances. The advantages are that if proper care is given the trees during the rebuilding period, much better formed trees may be obtained and a finer quality of bearing wood be secured throughout the new trees.

Fig. 16.—Same trees as Fig. 14, showing three seasons' growth after deheading. Photo taken January, 1917. No pruning done this winter. Trees have not borne more than a few almonds but have a good set of fruit buds.

To be successful, the deheading process must be followed by careful thinning of the sprouts that will be forced from the old stubs. If too much thinning is done the first year, there is danger of those that are left becoming top-heavy and breaking off. This is especially troublesome in sections subject to strong winds. The new sprouts should be left sufficiently thick to help support each other, and yet should be thinned out sufficiently to prevent crowding. At the end of the first season's growth, further trimming should be done, giving those branches best adapted for the frame of the new tree a chance to take the lead the following season. A sufficient amount of secondary growth should be left to help take care of the excessive sap flow in the shortened tree, but this must be watched to see that it does not interfere with the growth which is to be permanent. By the end of the second year, the permanent branches should be sufficiently strong and firm to permit the entire removal of all other branches. The amount of wood to be left until the end of the second season and then removed is clearly shown in figures 18 and 19. The third year the tree should make sufficient growth to restore very largely the equilibrium between the top and the roots and also produce a small crop. Figures 13 to 17, inclusive, illustrate the steps in rejuvenating an orchard by the deheading process from the start until the trees are again in commercial bearing.

Fig. 17.—Same trees as Fig. 14, showing four seasons' growth after deheading. Photo taken December, 1917, after trees have borne their first good crop of nuts. No pruning has been done since January, 1916.

Some growers find it more desirable to cut back only one-half of each tree at a time. The shock to the tree is not so great and the liability to sour-sap correspondingly less. The halves of the tree left standing act as a wind-break to prevent the blowing out of the new growth, which at the same time is less likely to break off because of its slower and more sturdy growth. The objection to this method is that the other half must be cut off the following winter and the shaping process is more or less uneven, and necessarily continued over a long time.

Fig. 18.—Two seasons' growth before pruning of almond trees deheaded six feet from the ground January, 1914, in the Armstrong orchard near Davis. The tree was only moderately thinned at the end of the first year. Photo taken February 2, 1916.

The other method of rejuvenation is to thin out gradually the dead and weakened branches in the old trees and rebuild them gradually, removing only a small portion in any one year. As the new growth is forced out it is used to replace the old wood where possible. Such a system requires more cutting of small branches high up in the tree to be successful, for it must not be opened to sunlight too suddenly or sunburn will result. The cutting of a large number of small branches tends to force the growth of new buds over a large proportion of the tree and if done moderately many of these will quickly develop into fruiting wood. Eventually, many large branches may have to be removed as newer branches are developed, but this must necessarily be a comparatively gradual process. The advantages of this method are that there is no sudden shock to the tree, there is always sufficient leaf surface to care for any extra supply of sap which may be forced into the tree by unfavorable weather and moisture conditions, and cropping will be continuous.

Fig. 19.—Same tree as Fig. 18 after pruning. The actual time required to prune this tree was twenty-five minutes. Photo taken February, 1916.

Whatever method of rejuvenation is used, the grower must be very careful from the first to protect all large wounds by some paint or other protective covering, such as asphaltum, to prevent checking and weathering and to keep decay from getting started and working into the heart of the tree. With very large wounds a protective covering must be kept on during the remainder of the life of the tree, or until the wounds heal over completely.

INFLUENCE OF CULTURE ON NUTS

The effect of culture on the nuts is quite marked. All varieties will vary in size from season to season and in different orchards during the same season, depending on the plant food and moisture supply available during the time the nuts are maturing. Some varieties, notably the Nonpareil, will vary more in size than in plumpness, while other varieties, like the Ne Plus Ultra, are more inclined to produce somewhat shriveled kernels and imperfectly developed shells and the variation in size will not be as great. As the trees bear larger crops the nuts produced are inclined to be smaller. Young, vigorous trees with a light crop will often produce unusually large nuts with comparatively thin shells. The value of a variety, therefore, will depend largely upon its behavior after the trees reach full maturity and bearing.

ORCHARD HEATING FOR FROST PREVENTION

There has been considerable interest in protecting almonds from frost because of the frequency with which they are subjected to such conditions by reason of their early blossoming habit. The use of orchard heaters in some districts has become a common and successful practice. It is possible to economically control several degrees of frost for three or four nights or possibly more by a judicious use of heaters.

The best type of heater for almond orchards has not been thoroughly worked out as yet, but the reservoir heaters of the Bolton or Hamilton type have been commonly and successfully used. The former are commonly known as "pots." At least 75 one or two-gallon heaters per acre are necessary to control temperatures as low as 27 or 28 degrees F. Probably not over 100 pots to an acre would be needed at most. Double the number of pots should be placed around the outside row of the orchard to afford the necessary protection.

The gravity of oil best suited for orchard heating is between 20 and 25 degrees Beaumé. It is essential that it be as free from impurities, such as sulphur, as possible. During the spring of 1917, orchard heating was done very extensively in some districts of the state while the trees were approaching full bloom. As the season progressed it became apparent that some serious injury had been done by the heating, for the bulk of the blossoms fell off and the leaves turned yellow in streaks as if burned by an acid. The appearance was exactly as if the particles of soot settling on the pistils of the flowers and on the young leaves had absorbed sulphur dioxide gas (a product of oil combustion where sulphur is contained in the oil) and that the dews uniting with it to form sulphurous acid had done the damage. Had the heating been done later and only after the trees had passed full bloom, it is possible that a much heavier set of nuts might have been secured, since the small fruits, where they had formed, seemed to have been uninjured, only the pistils of the flowers having been affected, probably preventing pollination and fertilization of the ovules.

The time of heating is therefore a very important point. Almond blossoms become progressively more tender to frost as they advance in development. With their petals on they are not nearly as tender as they are after the petals have fallen. They reach their most tender stage after the calyx shucks have fallen from the young expanding fruits before they are the size of a pea. Ordinarily, orchard heating before the trees have passed full bloom is a waste of time and material and is often injurious. The most needed time is for a period of two or three weeks after the bulk of the petals have fallen, unless temperatures below 28 or 29 degrees are encountered before that time.

CROP HANDLING

Harvesting.—The harvesting of the crop should be commenced as soon as the hulls have opened to their fullest extent and no time should be lost in completing the work. The nuts in the center of the tree are the last ones to ripen and so may be used as indicators. If harvesting is commenced early, the nuts will cling to the tree rather tenaciously and knocking must be very vigorous in order to shake them loose. On the other hand, if they are allowed to hang too long after ripening, a number of difficulties may be encountered. They may be blown to the ground by light winds and the cost of gathering be increased, as quite commonly occurs with the Peerless, or the hulls may dry up and in doing so, close around the nuts to a greater or less extent and add to the cost and difficulty of hulling. This is most noticeable with the Nonpareil. Strong winds will break off a great many of the nuts of any of the varieties, and promptness is doubly essential where there is any likelihood of such winds during the harvest season. Depredations by birds may cause serious losses, especially with the soft and papershell varieties. Infestation by worms may often be quite serious in the papershell varieties when they are allowed to hang too long. In case of damp or foggy weather the shells turn dark and sometimes commence to mildew, requiring heavier bleaching to brighten them sufficiently for market demands. Rain stains can never be removed entirely by bleaching.

Fig. 20.—Harvesting almonds by knocking onto sheets spread on ground.

The crop is gathered by knocking the ripened nuts and hulls with long poles onto sheets spread on the ground under the trees. The knocking should be done near the portions of the trees where the nuts are borne and by striking a number of light, quick blows, rather than by a heavy blow aimed to jar a large branch. This will avoid injury to the bark by bruising and will accomplish the work in less time. The blows should always be delivered squarely against the branch. A glancing blow will tear the bark and break off a great many fruit spurs, thus reducing the bearing surface for the next year. For this reason harvesters must be watched closely all the time to insure the proper use of the poles. The sheets, two in number, are spread under the trees so they will overlap and catch all the nuts that fall (see figure 20). When sufficiently loaded with nuts to make dragging the sheets from one tree to another difficult, they are emptied into lug boxes and sent to the huller.

The character of the harvesting equipment may vary considerably, depending on the acreage, character of trees and ground, time required, capital available, and the personal wishes of the owner. Some growers use heavy poles of pine, spruce or fir, while others prefer the lighter bamboo poles. The heavy poles are from 1½ to 2 inches in diameter at the base and from ¾ to 1 inch in diameter at the top. Experience has shown that poles of this type over 20 feet in length become unwieldy, and not only swing slowly but do more damage because of the greater difficulty of control. Most growers prefer 16-foot poles with a few 20-foot ones for use in the tops of the taller trees. Where the tops cannot be reached with these, the men climb into the trees with shorter poles. The bamboo poles used are about 24 feet long and because of their lightness can be used with greater speed. Bamboo poles with short internodes should be selected as they are less likely to break. Breakage may also be reduced by storing the poles in a cool place where drying-out will not be excessive.

The sheets used are made of duck ranging in weight from 7 to 12 ounces. Sizes of single sheets range from 12 × 24 to 24 × 48 feet. Sheets need not be much longer than the longest diameter of the tree. To prevent mildew and rotting of the fabric in the sheets they should be boiled in a solution of tannin before being used. The life of sheets thus treated will be greatly lengthened.

A number of growers have provided special contrivances by which sheets are mounted on sleds or wheels so they are not dragged on the ground. The principal objection to such an arrangement is that the sled or wheeled frame must be made in two sections, one for each side of the tree, thus increasing the cost of the operation because the horses can be used for nothing else while harvesting is in progress. On the other hand, by the dragging process sheets will not last more than two or three seasons, whereas by the sled or wagon method they will last from six to ten years longer. The two wagons shown in figure 21, each 12 × 24 feet, cost between $60 and $70, about 1914. The canvas portion is of 8-ounce duck. By this method the knockers can gather ten lug boxes before emptying. The work can be done much faster with wagons. With the sleds a sheet is fastened lengthwise on the right-hand side of one and another on the left-hand side of the other sled.

Hulling.—After harvesting, the almonds, hulls and all are taken while still moist, to the huller to separate them. If they become dry before hulling they must be dipped in water or the shells will be broken. All hulling was formerly done by hand, and this is still done where only small lots are to be handled. The invention of machinery for this purpose has reduced the cost of hulling from 60 to 80 per cent, for most outfits separate the hulls from the nuts before they leave the machine. Some of the hand hullers consist simply of the hulling portion of the large power machines without the separating screens. These cost about one-sixth or one-seventh as much as the large machines, and where a man has a small acreage and is too far away to haul to a large huller, a hand machine will greatly facilitate the work, even though the final separation must be done by hand.

Fig. 21.—Portable almond sheets mounted on wheels as used by N. J. Lund, Oakdale, California, 1916.

There are three different kinds of hulling and separating machines now in operation in California, all invented by California almond growers. The first one made was the Read "Sure-Pop" almond huller. This is now manufactured in three sizes by the Schmeiser Manufacturing Company, Davis, California. The No. 3 huller does not have any separating device and is generally best for orchards of less than ten acres. It may be operated by hand or by a small engine or motor. The No. 2 hullers both hull and separate and are operated only by power. They should pay in orchards of ten acres or more. The No. 1 is the largest made and is for use in large orchards of 100 acres or more.

The Beach huller is of more recent origin, having been in use only since 1895. It was invented by J. E. Beach of Fairoaks, California, and is being manufactured by him. The two sizes of this machine are both power outfits; they are doing satisfactory work at the present time.

The third huller is that made by C. U. Reams of Suisun. One of the first machines made by him was in 1897, and is now in working order at the F. O. Scarlett ranch, northeast of Suisun, and is doing satisfactory work. Since the first invention, Reams has made a number of improvements both in the method of hulling and of separating.

Many growers do not have sufficient tonnage to enable them to afford a commercial huller, and yet hulling by hand is a slow and tedious practice. C. E. Sedgwick, Manager of the Solano District of the Pacific Gas and Electric Company, located at Dixon, California, conceived the idea of using a centrifugal blower operated by a small electric motor to do the hulling. His description of this outfit, quoted from "Pacific Service Magazine," April, 1916, page 393, after making two small corrections given by Mr. Sedgwick, is as follows:

The equipment consists of a No. 0 Sturtevant exhaust fan belted to a 1 h.p. motor. The nuts are fed into the suction side of the fan where they are picked up by the runner, hurled against the casing of the fan and blown out of the discharge into a box.

The motor consumes three-tenths kilowatts when almonds are fed into the fan at the rate of a lug box every minutes and one-half, so that the power cost, even at the 8-cent lighting rate, is only 2.4 cents per hour. The fan costs about $20, while the regular commercial hullers run as high as $750.

Further inquiry from Mr. Sedgwick developed the fact that this huller has operated for three seasons on a 20-acre almond orchard. Peerless, Drake and I.X.L. almonds were all hulled successfully. Nonpareils have not as yet been tried. The speed most commonly used was about 1200 r.p.m., though it varied somewhat with the different varieties. He believes that a larger size would do better work.

The efficiency of any of the above hullers depends largely upon the speed of the machine and upon the condition of the almonds in the hulls. If the machine runs too fast the almonds will be broken and injured, and if it runs too slowly many of the almonds are not hulled. If the almonds are allowed to hang on the trees too long, or if allowed to lie around too long after being harvested, the hulls become dry and leathery and the difficulty of hulling is greatly increased. Dipping in water in such cases may help to overcome this difficulty to a limited extent.

The papershell varieties, notably the Nonpareil, are much more difficult to hull without breaking the shells than are the harder shelled varieties. The Nonpareil hull has a tendency to close around the nut on drying, making hulling under such circumstances very difficult.

Often when the moisture supply in the soil is exhausted before the nuts are ripe, or where the loss of leaf surface due to mites is serious prior to ripening, the hulls open only slightly or not at all, but dry onto the shell of the nut. Such "sticktights" can only be disposed of profitably by allowing them to thoroughly dry, when they are cracked and sold as kernels.

After the hulling operation all almonds must be gone over by hand to remove pieces of hulls and inferior or gummy nuts. Where canvas drapers are not available for sorting directly from the huller, the nuts are piled in hoppers and sorted on benches beneath them.

Drying.—Immediately following the sorting, the nuts are spread on trays and thoroughly dried in the sun. In the interior valleys during the hot, dry weather the nuts will sometimes dry so quickly that by the time the sorting from hoppers is completed the nuts are sufficiently dry to be bleached. The grower must be certain, however, that such is the case before any bleaching is done, or before the almonds are delivered to the warehouses for bleaching. The nuts are sufficiently dry when the kernels will break without bending. Quick drying is essential to prevent the excessive darkening of the shell.

Bleaching.—When thoroughly dry the nuts are ready for bleaching. The shells are first moistened by spraying with water or subjecting them to low-pressure steam for 10 to 20 minutes. The shells are then subjected to the fumes of burning sulphur for 10 to 30 minutes. The sulphur fumes are absorbed by the moisture on the shells, which are bleached to a bright yellow color. After bleaching the nuts are exposed to the air for a few moments to allow them to dry. The market demands a nut that has been sufficiently bleached to give it a bright, clean, yellow color. An over-bleached almond is equally objectionable because of its pale, sickly, yellow or whitish color. Over-steaming or sulphuring permits excessive penetration of the sulphur fumes, with the resulting danger of absorption by the kernel. While this may not be noticeable in the flavor, it will eventually result in premature deterioration in the form of rancidity. Unbleached almonds remain edible much longer than bleached almonds in nearly every case. Ordinarily one to three pounds of the best flowers of sulphur is required to bleach a ton of almonds. Lump sulphur is not satisfactory.

Sacking.—During the preliminary handling of almonds ordinary grain sacks are commonly used. After bleaching, in which condition they are ready for market, they are put in standard almond bags, measuring 20 × 40 inches and weighing 1¼ pounds. The weight of a bag of almonds will vary, depending not only on the variety but also on the year in which the crop was grown and the locality in which it was produced. For selling purposes the California Almond Growers' Exchange estimates weights of different varieties to be as follows: Nonpareil about 85 pounds to the bag; I.X.L. about 80 pounds; Ne Plus Ultra, 75 pounds; Drake, 90 to 100 pounds; Languedoc, 100 pounds; and hardshell almonds, 100 to 120 pounds.

Shelling.—Within the past two years the shelling of almonds has taken a prominent place in the consideration of the men charged with the disposal of the almond crop. The increasing popularity of shelled almonds, and the limited market for unshelled almonds, makes the production of more shelled almonds imperative in view of the prospects of greatly increased production in the next few years. A small proportion of the shelled almonds marketed are those accidentally shelled during the hulling process. This probably averages less than 30 pounds per ton of almonds hulled in ordinary years.

The varieties most commonly shelled are the papershells. They are much more easily shelled without breaking the kernels than are the harder shelled varieties and, in addition, are worth more for shelling because of the high percentage of kernel compared to shell.

Grading.—Grading almonds for size is not done at present but probably will be within the next few years. Grading for quality is done regularly by testing an entire lot rather than attempting to separate inferior nuts. The standard grade consists of all lots having the required percentage of good kernels, free from worms or gummy nuts. This requirement varies between 90 and 95 per cent, depending on the condition of the crop as a whole and on the market conditions. The standard grade of a given variety sells on guarantee that it shall be up to advertised standard. All lots which cannot pass this are sold on sample, and therefore, on their own individual merits.

MARKETING

The marketing of the California almond crop is at present on a firmer basis than at any time in the past. Previous to 1910 there was little or no coöperation among growers and the buyers had everything their own way. In May of that year, however, Mr. J. P. Dargitz, an almond grower near Acampo, California, successfully organized the California Almond Growers' Exchange, consisting of nine local associations with a total membership of 230 growers. The Exchange started business with $1000 borrowed capital, personally guaranteed by the directors. On June 1, 1918, there were 22 sub-associations representing about 2000 growers, controlling about three-fourths of the crop. The Exchange now is not only out of debt but owns investments aggregating $100,000 in value, including warehouses, a central shelling plant and other property. At the same time, the growers have been receiving about 50 per cent more for their almonds than before the Exchange was organized.

The success of the Exchange, with the consequent higher prices to the grower, has resulted in a large increase in the acreage of almonds in California. This increase is making it necessary to develop new markets to absorb the greater tonnage, and this can only be done effectively or satisfactorily by coöperative effort.

Heretofore, California almonds have been marketed chiefly in the shell. A small proportion has been cracked annually to supply western brokers and confectioners, and practically all of these have been sold west of the Rocky Mountains. Only the whole nuts have been shipped to the eastern markets.

The European crops are sent to the United States largely as kernels and have had a practical monopoly of the shelled almond business east of the Rockies. The Tarragonas and Valencias shipped to America come into direct competition with the I.X.L., Drake, Languedoc and other California almonds, all of which are unshelled for the eastern markets. The imports of unshelled almonds average about the same as the California production. The Jordan, Princess and other varieties, however, come in shelled, constituting the bulk of the importations. The Jordan, because of its superior quality, is in a class by itself and does not compete at present with the California product. The Princess and other almonds of that type are much the same as the California shelling varieties and will be serious competitors when sold in the same markets.[4]

Storing.—Almonds awaiting removal or sale will become rancid if stored in warm or damp places. If the almonds have been properly cared for during the handling process to prevent worm infection, and if the nuts have been thoroughly cured, they will keep satisfactorily for a year or more. The ideal condition is to keep them stored in a uniformly cool, dry storage place with ample ventilation.

YIELDS

The yield of almonds in different years and in different orchards is probably more variable than that of any other of the common orchard fruits. The fluctuations from year to year are largely due to climatic conditions, while the variations in different orchards are largely due to variety, care given the trees throughout their life, the character of the soil, and location with relation to local frost conditions.

Almonds first commence bearing at from two to four years of age; the first crop ranging from one or two nuts up to a hatful or possibly more. The trees will first commence to bear a crop which it will pay to harvest, at from three to five years of age. Ordinarily, it will be nearer the latter, depending upon the type of soil in which the trees are growing and the moisture conditions surrounding them. On the hill lands the trees begin to mature much earlier than in the rich bottom lands and consequently come into bearing earlier. It must be borne in mind that a crop which it will pay to harvest does not necessarily pay for the cost of orchard maintenance. A crop is not considered a paying crop until it pays for the cost of maintenance as well as harvesting and handling. Almond orchards, as a rule, reach this point at from five to seven years of age. From this time on the trees should continue to increase in production from year to year, allowing for failures due to frost and other unfavorable conditions, until they are from 12 to 20 years old. Under the common methods of care that most orchards receive, the trees commence to decrease in their production at from 25 to 30 years, although in some cases it will be even sooner than that. On the other hand, well cared-for orchards will continue their maximum production even longer. The age at which an orchard will no longer pay will range from 30 years upward. The top limit is still unknown.

Investigations carried on during 1913-1914[5] brought out the following facts: The average production of almonds in California is between 700 and 800 pounds per acre; if care is exercised in the selection of a proper location for an orchard and if good judgment is used in managing it, 1000 pounds per acre would be a safe estimate for business purposes; in many years competent men might be expected to obtain 1500 pounds per acre, but this could not be expected to hold for a ten-year average. The possibilities are shown by the crop from one acre on the University Farm, at Davis, California, of ten-year-old trees which amounted to nearly 2800 pounds in 1917.

COST OF PRODUCTION

The cost of producing almonds involves a number of variable factors, including overhead charges, such as the cost of the land, equipment, taxes, insurance and depreciation; and also the cost of maintenance and handling. Maintenance includes such costs as pruning, plowing, cultivating, spraying and irrigation. Handling includes harvesting, hulling, hauling, and warehousing. Tabulations of estimates in tables II to VIII are based on information collected during the years 1913 to 1916, inclusive, from a large number of growers in practically all the almond districts of the state, and represent as accurately as possible with the data at hand the average costs which actually exist throughout California.

Cost of the Land.—The estimates given in table II are to be taken as only partially indicative of conditions which actually exist in the various districts mentioned. These figures do not give the entire range of prices but indicate some of the more common values placed upon the land.

TABLE II
VALUE OF ALMOND LAND IN CALIFORNIA

District—	Bare land	Land in bearing orchard
Best Sacramento Valley land	$200-$400	$400-$600
Other good interior valley lands	150- 300	400- 500
Sacramento Valley, foothill sections	75- 150	200- 400
Contra Costa County	100- 300	250- 500
Santa Clara and San Benito counties	300- 600	500- 800
Paso Robles district	50- 150	....
Banning district	400- 800	600-1000
Averages for California	$250	$500

Equipment.—It has been found impossible to gather accurate figures upon cost of equipment in almond orchards, and especially so in view of the present abnormal economic conditions, but the list given in table III will give an idea of the equipment required. In addition, there will be other small items the grower will need which are not mentioned here.

TABLE III
ALMOND ORCHARD EQUIPMENT

Plows
Harrows (spike-tooth and spring-tooth)
Disc Cultivator
Weed cutter
Clod masher
Roller
Hoes, shovels, etc.
Pruning tools
Brush burner
Spray outfit
Wagon
Barns, sheds and other buildings

Harvesting equipment:
Almond sheets
Poles
Lug boxes
Hulling machine
Sorting tables and bins
Drying trays
Sacks for transportation to warehouse
Orchard heating equipment ($25-$30 per acre)
Horses or tractors
Harness

Average Overhead Charges.—Table IV shows the average overhead charges for almond orchards. Interest and depreciation on buildings are not included because of the great variation in their character, so that an extra charge must be figured on these items by the individual grower.

TABLE IV
AVERAGE OVERHEAD CHARGES PER ACRE

Taxes and insurance	$4.00
Interest	30.00
Depreciation on working equipment	4.00
Total	$38.00

Cost of Production.—Table V shows the average cost of production for bearing orchards of varieties in all districts, based on personal observation and cost records from a large number of orchards mentioned previously. Wherever cost is dependent upon tonnage the average yield of 700 pounds per acre is used as the basis for computation. Depreciation on buildings and trees, time spent by teams in idleness, feed consumed during such times and other minor items are too variable to safely estimate, but must be considered.

TABLE V
AVERAGE COST OF PRODUCTION OF ALL BEARING ORCHARDS IN CALIFORNIA

	Per acre	
Maintenance:		
Pruning	$3.00	
Plowing	2.75	
Harrowing	.75	
Cultivation and weed cutting	3.00	
Spraying	3.00	
Irrigation	2.00	
Handling:		
Harvesting, hulling, etc.	20.00	
Warehousing (including bleaching), @ ¼c per lb.	1.75	
Miscellaneous expense for maintenance and handling	2.00	$38.25
Overhead charges		38.00
Total cost per acre		$76.25
Cost per pound for maintenance and handling		$0.055
Cost per pound for overhead charges		0.054
Total cost per pound		$0.109

Returns.—Prices paid to growers have fluctuated considerably, due to the great variation in both the California and European crops from year to year. Table VI shows the average prices per pound paid to the growers for the four principal varieties marketed through the Exchange since its organization.

TABLE VI
NET PRICES REALIZED BY THE EXCHANGE MEMBERS FOR DIFFERENT VARIETIES FOR THE YEARS 1910 TO 1916, INCLUSIVE, IN CENTS PER POUND

Year	Nonpareil	I.X.L.	Ne Plus Ultra	Drake	Crop tons
1910	14.00	13.00	12.00	10.00	3,500
1911	16.50	15.50	14.50	12.00	1,450
1912	13.25	12.25	11.25	9.50	3,000
1913	17.25	16.25	15.25	13.25	1,100
1914	18.00	15.00	14.50	11.50	2,250
1915	13.00	12.00	11.00	9.25	3,500
1916	17.25	14.75	13.75	13.00	3,400
Average	15.61	14.11	13.18	11.22	2,571.4

Table VII shows the average price per pound paid to the growers for all almonds (unshelled) regardless of quality and variety, based upon the entire crop handled by the Exchange during the years 1910 to 1916, inclusive. From these figures the average return per pound for all varieties for seven years based on the crop tonnage for each year, 1910 to 1916, inclusive, has been found to be 13.09 cents per pound.

TABLE VII
AVERAGE PRICES PER POUND PAID GROWERS FOR ALL UNSHELLED ALMONDS FOR THE YEARS 1910 TO 1916, INCLUSIVE

Year	Price per pound, cents	California crop, tons
1910	12.0	3,300
1911	13.5	1,450
1912	11.0	3,000
1913	15.5	1,100
1914	14.05	2,250
1915	10.75	3,500
1916	13.97	3,400

| Average | 13.09 | 2,571.4 |

The relation of yields, returns and profits from the growers' standpoint is one which every person must consider before entering the business. In view of the extravagant claims which have been made as to the enormous profits realized by the average grower, the figures in tables II to VIII have been worked out and presented here. The summation of the relation of yields, returns and cost of production to the profits for the average grower of almonds is shown in table VIII.

TABLE VIII
RELATION OF AVERAGE YIELDS, COSTS AND RETURNS, TO PROFITS

Average yield per acre	700 pounds	
Average returns to grower per acre		$97.30
Average cost of production per acre		76.25
Average profit per acre		$21.05

Depreciation on buildings and trees, and other unfigured costs, are too variable to estimate, but they must come from these profits.

DISEASES

Crown Gall.—Also commonly known as root-knot. This disease is one of the most serious with which the grower has to contend. It is found practically everywhere almonds are grown and either greatly reduces the vitality of or kills the trees affected, depending upon the seriousness of the attack.

The disease is caused by a bacterial organism, *Bacterium tumefaciens*, that seems to be native to most California soils. It is characterized by large swellings on the root crown or main roots just below the surface of the ground, though lesser infections may sometimes be found also on the smaller roots. When cut open, these knots appear spongy as if the bark and wood were all mixed together in one mass. They are most serious when spread over a large surface, either partially or completely girdling the root or crown of the tree.

Control methods are of three kinds:

(1) Plant nothing but clean, healthy nursery trees, free from all trace of galls. In planting these trees be careful to trim off all broken or injured roots, leaving nothing but smooth clean cuts at the ends of the roots which will heal over readily with the minimum opportunity for infection.

(2) Galls on orchard trees may be cut out to clean, healthy wood with a sharp knife or gouge chisel. The wound should be thoroughly disinfected with a strong copper-sulphate or corrosive-sublimate solution, and painted with a protective covering such as paint or melted asphaltum, or it may be covered

directly with Bordeaux paste and then the earth returned to its place over the roots.

(3) A method used with apparent success is to bore a one-inch hole about two-third of the way through each gall, as soon as the trees have become dormant in the fall. Then fill each hole with a concentrated solution of copper sulphate and plug the opening. By spring, when growth is ready to start, the gall may be knocked off with a hammer. In most cases the gall is so thoroughly permeated by the solution that the infection is completely killed and further gall growth ceases in that place, unless later infection occurs.

The use of resistant stocks has been suggested as a means of avoiding infection, but no such stock suitable for the almond has yet been proved to be sufficiently resistant under average conditions to be safely recommended. The greatest hopes for future success in combating this disease, however, lie along this line.

Oak Fungus.—This is one of the most difficult diseases to control because it works and spreads beneath the surface of the ground in the roots of trees. In some sections of the state it is very serious in many orchards.

The disease is often known as root-rot, being caused by a fungus *Armillaria mellea*, commonly called "toadstool" fungus. It is known as Oak fungus because the disease is most commonly found in spots where old oak trees have stood. Where orchards have been planted on such land, spots appear in which the trees gradually die, the disease spreading from tree to tree, in ever-widening circles, involving ordinarily about one row of trees each year. During the winter, clusters of toadstools may be seen at the base of the affected trees. The fungus lives over in the old oak roots for many years and, as the orchard becomes well established, the fungus spreads to the almond roots. If not checked the spot will eventually involve the entire orchard and prevent further growth of almonds on such land for many years.

Control is very difficult but may be secured by digging a deep trench around the affected area and preventing the infection from passing beyond through the roots. The spread of the disease may sometimes be held in check by grubbing out a row or two of healthy trees outside the affected area and taking care that all of the large roots are removed to a depth of several feet. Carbon bisulphide has been suggested for killing the fungus, but the cost is prohibitive except in small spots just starting.

There are no resistant stocks known at present upon which the almond can be worked. The fig, pear or black walnut might safely replace the almond in such spots.

Shot-hole Fungus.—There are three different fungi that produce the shot-hole effect on the leaves of the almond, thus giving rise to the name.

(1) *Coryneum beyerinikii*, or peach blight, is the most common form. It is not as serious on the wood of the almond as it is on the peach, but in seasons of damp spring weather it does much damage to the blossoms, fruits and leaves. Affected blossoms are killed outright, the entire blossoms turning brown and dropping much as if killed by frost. The young fruit becomes spotted by the fungus and this causes malformation, gumming and shriveling of the nuts, varying considerably with the severity of the attack. On the leaves many small dead spots appear, the dead tissue soon falling out and giving the shot-hole effect. Where the twigs are affected, small dead spots appear during the winter, most often at the buds. This causes the death of the buds and often the ends of the twigs. During the spring, after growth starts, considerable gumming occurs from these spots.

Effective control can only be secured by two sprays—Bordeaux mixture in the fall, as soon as the tree becomes dormant, and either Bordeaux or lime-sulphur solution (winter strength) just before the buds open in the spring. Both fall and spring sprays must be thoroughly applied to be effective.

(2) *Cercospora circumscissa* is another fungus causing much the same effect as the Coryneum. It is difficult for an untrained person to distinguish between them. The same sprays used for Coryneum are effective in controlling this, though if this form alone is present the Bordeaux mixture or lime-sulphur spray in the spring should be sufficient.

(3) *Gloeosporium amygdalinum*, while apparently uncommon in this state, has been found to exist in some places. Further work must be done on this to determine its behavior and the most satisfactory methods of control, but it is believed that the control measures mentioned for the other forms of "shot-hole" will also be applicable to this.

Prune Rust (Puccinia Pruni).—This fungus is worst in the southern coast sections where almonds are not extensively grown. It is not serious on thrifty trees well supplied with moisture. The disease is characterized by reddish pustules on the under-sides of the leaves, appearing generally about July or August and causing a premature yellowing and dropping of the leaves.

Ordinarily the only treatment needed is to supply the necessary moisture in the soil to keep the trees healthy and vigorous.

Heart Rot.—This is one of the most insidious of tree diseases, for it works inside beneath an apparently healthy exterior until the decay has progressed so far that the tree commences to break down, and then it is too late for remedial measures. The almond is not as susceptible to this as most other kinds of orchard trees, but where large wounds have been exposed to the weather, infection may take place readily, and after it is once well started it continues at a comparatively rapid rate. Decay is caused in most cases by one or more of about a dozen different fungi, of which the oyster-shell fungus is by far the most common.

Control consists in taking care to leave no open wounds exposed to the air to dry and crack, thus permitting the entrance of decay organisms. Much of this can be avoided by care in pruning the young tree so that the removal of the large limbs will not be necessary later on. Where such wounds must be made, measures should be taken to prevent infection. This can best be done by making smooth, clean cuts close to the part from which the branch to be removed emanates, leaving no stub. Stubs dry out and crack more quickly and require very much longer to heal over, if this is possible at all. Further, all such wounds which will not heal over the first season should be covered with some good disinfectant, such as corrosive sublimate, one part to one thousand parts of water, and then painted over with some elastic coating, such as "Flotine" or asphaltum, grade D, applied with a brush. The entire wound must be covered or the work is largely wasted.

Die-back.—This is serious in many orchards where moisture is insufficient to carry the trees through the growing season, and the trees show considerable dying-back of the branches. Unfavorable soil conditions, such as hardpan, gravel or sand may be the direct cause of such moisture shortage. Lack of soil fertility is also a common cause. Control measures consist in remedying the defective conditions and where this cannot be done economically it is better to abandon further attempts at almond culture on such land.

Sour-sap.—This is one of the so-called "physiological diseases" and is quite common with the almond. It is most frequently found where trees are planted in heavy or poorly drained soils. The inability of the almond to endure standing water around its roots for any length of time and particularly so after growth commences in the spring, renders it especially liable to sour-sap when planted in soils where excess water from the late winter and spring rains cannot be readily drained away. The direct cause of the trouble is sudden changes in weather from warm to cold after growth commences, which checks the flow of sap very suddenly, causing stagnation, cracking of the bark and then fermentation. With an unusually strong flow of sap in trees in wet soils, such climatic changes cause unusually severe disturbances in the normal functioning of the trees.

The affected trees ordinarily show the disease first in the spring when gum may be seen oozing from the bark of the trunk or main branches, and sometimes even from the smaller branches. Small or large branches may die, and in severe cases the tree may die soon after having commenced to leaf out strongly. On cutting through the bark to the wood and peeling back, a strong sour odor is noticeable. The cambium layer appears brownish or reddish in color and often masses of gum may be found between the bark and the wood. Mild cases may not be serious enough to show on the outside of the tree and only portions of the cambium layer may die. The sudden dropping of the blossoms or young fruit may in some instances be attributable to sour-sap.

All affected parts on smaller branches should be cut back to healthy wood, while on the main branches or trunk, where only a small portion or one side is affected, it is best to clean out the dead bark and paint the bared wood with a protective covering until new bark can cover the spot. At the same time every effort should be made to remedy the soil-moisture conditions which were largely responsible for the trouble in the first place.

Fruit-drop.—The same conditions which cause sour-sap may cause fruit drop. It may be caused by lack of pollination due to improper mixing of varieties or to rain during blossoming. Frost may also produce the same thing by killing the germ in the young fruit. In such cases, the fruit may remain on the tree for one or two weeks after the injury occurs before falling, and in some cases, may even appear to continue its development for a short while.

INSECT PESTS

Mites.—Commonly called red spiders. There are two kinds of mites that do much damage in almond orchards, the brown or almond mite and the yellow or two-spotted mite. Both are common in all parts of the state and are the worst pests the almond grower must regularly face.

The brown mite (*Bryobia pratensis*) is the larger of the two, is dark red or brown in adult stage, has very long front legs, and a flattened back. It does not spin any web and works on the green bark of the small twigs as well as on the leaves, sucking the plant juices from beneath the bark. It causes a mottling of the leaves which eventually fall, although not as readily or in such large numbers as when attacked by the yellow mite. The injury to the tree is equally as great because of the serious drain on the vitality as a whole and because it commences work earlier in the season. This mite may spend its entire life on the tree; the very small, round, red eggs being laid largely on the under-side of the branches and in cracks and crevices in the bark and twigs. These remain on the tree throughout the winter and hatch early in the spring soon after the trees have their leaves half developed, leaving the white egg-shells in place. Most of their work is done in the spring and early summer.

Fig. 22.—Nonpareil almonds. Branch on left free from Red Spider and holding its full supply of leaves in green, healthy condition; branch on right defoliated by Yellow Mite. Note premature ripening of nuts on defoliated branch.

Fig. 23.—Forty-three-year-old Languedoc almonds on peach root, near Davis. This orchard has not been thoroughly cultivated or irrigated. Mites have largely defoliated the trees. See contrast in Fig. 24.

The brown mite may be controlled satisfactorily by means of a dormant spray of lime-sulphur solution, 1 gallon to 10 gallons of water, applied just before the buds open in the spring. Crude oil emulsion as a dormant spray is also effective, if thoroughly applied over the entire tree under high pressure. This also applies to applications of the lime-sulphur spray.

During the growing season a milder material must be used. Dry dust sulphur, using only the very finest grade of "flowers of sulphur," is often very effective, provided weather conditions are satisfactory, but generally this must be applied a number of times if best results are to be obtained. The work is done by blowing the sulphur dust into the tree with blowers in the early morning when there is little or no wind.

A more satisfactory method is the use of "Atomic sulphur" or other sulphur pastes or similar material. "Atomic sulphur" is a prepared spray whose value consists in the fact that the sulphur is held in suspension in water so that it may be applied as a liquid spray. By this method the material may be more effectively and thoroughly applied. "Atomic sulphur" is applied at the rate of 10 pounds to 100 gallons of water.

Fig. 24.—Forty-three-year-old Languedoc almonds on peach root, near Davis. This orchard has been irrigated and thoroughly cultivated, and mites have not defoliated the trees. Note contrast in Fig. 23.

The use of lime-sulphur, the commercial strength of 33 to 34 degrees Beaumé being diluted 1 part to 35 parts water, is another effective method.

For such use a flour paste may be added at the rate of 4 gallons to each 100 gallons of the spray mixture to act as a spreader. This paste is made by cooking one pound of flour with enough water to make one gallon of the mixture.

The yellow mite (*Tetranychus telarius*) is much smaller than the brown mite and is of a pale yellow color with occasionally a reddish tinge and sometimes with two darker spots on either side of the body. Unlike the brown mite, the winter is spent in concealment somewhere, presumably off the tree. During the warm days of early summer, generally in June, the mite makes its appearance on the trees, spinning a fine web on the leaves, generally on the upper surface, and then works under this web. The mite sucks the plant juices from the leaves giving them a yellowish mottled appearance. These leaves soon die and drop to the ground. In serious infestations the trees are often almost completely defoliated by the end of August. (See figure 22.)

The use of dormant sprays is not effective for controlling the yellow mite, but the summer sprays mentioned above are all satisfactory, and for best results, must be applied under high pressure, preferably 200 pounds or more.

The mites are much easier controlled where a comparatively high percentage of moisture is kept in the soil by frequent cultivations, or, if necessary, by irrigation (figures 23 and 24).

Peach Twig Borer (Anarsia lineatella).—In the larval stage these borers work on the young buds and shoots in the early spring. They are especially troublesome in newly planted orchard trees when a comparatively small number may kill most of the new shoots which are needed to make the desired framework of the tree. In some years they may be serious in large trees also. Dormant spray of lime-sulphur applied under high pressure just as the buds are opening in the spring will control the borers very effectively. This same spray may be used to control the brown mite, thereby accomplishing double control.

California Peach Borer (Ægeria opalescens).—The larvae are serious in many parts of the state where they burrow just under the bark near the surface of the ground. They may be detected by the small bits of frass and gum at the entrance of their burrows. If allowed to continue, they will eventually girdle the tree.

The surest means of control is to dig out the worms with a knife or kill them with a wire probe. This work should be done systematically once or twice every year and very thoroughly if it is to be effective. The application of hot "Flotine" or asphaltum, grade D, after the worms are killed should help considerably to prevent the entrance of the larvae. To be effective it must be applied at least twice a year on young trees and probably the same on old trees.

Thrips.—These are most serious on the almond leaves, their attacks being serious enough to cause considerable defoliation in late spring or early summer. They may be controlled by spraying with lime-sulphur, 1 to 30, to which has been added black-leaf 40 (40 per cent nicotine) at the rate of 1 part to 1500 parts of water or other spray-mixture. It is possible that other than the pear thrips have been doing damage, but the same spray as described above should be effective against all.

Grasshopper.—Grasshoppers have been serious in orchards in outlying foothill districts in some years, and especially so in young orchards where it has been almost impossible to get trees started properly. In such locations special means must be employed on a large scale to protect orchards from their devastations, of which poisoned bait and hopper dozers are the most effective.

Fig. 25.—Almonds infested with larvae of Indian Meal Moth (*Plodia interpunctella*).

Indian Meal Moth (*Plodia interpunctella*).—The larva of this moth feeds on the kernels of the harvested almonds when they are stored. Infestation usually takes place in storerooms or warehouses in which the nuts have been placed for a time. The warehouses become infested from old grain bags which have been kept there at one time or another. So far as known, infestation does not take place in the field. The larvae will continue to work in the stored almonds for a long time, doing a very great amount of damage (fig. 25). They may be controlled by thoroughly cleaning out the corners of the warehouse and thoroughly disinfecting. The nuts should be disinfected with carbon bisulphide (explosive when in the form of a gas mixed with air), or other means used to control insects in grain.[6] Prevention is far easier than the cure in this case.

Scale, aphis, diabrotica and other insects are sometimes found on the trees, but are generally not sufficiently troublesome to require special attention. Most of them are held in check by the control measures used for the more serious pests.

OTHER PESTS

Gophers.—These often do great damage to the trees by girdling them just below the surface of the ground, or if they do not actually girdle the trees, they cut them enough to devitalize them and, in addition, the wounds made by their gnawings frequently become infected with crown-gall organisms. The only safe means of control is the constant use of traps supplemented by poisoned bait.

Squirrels.—Squirrels are very troublesome, as they harvest a large amount of almonds before they are sufficiently ripe to be harvested by the grower. The use of poisoned grain or "gas," if used over a large area of surrounding territory, will prevent serious depredations.

Birds.—Birds also carry off large amounts of almonds if the orchards are near open country or hills, especially if wooded. Crows, bluejays, blackbirds, yellow-hammers, robins and other similar birds are the worst offenders. Sometimes linnets eat off large numbers of fruit buds in the spring in a few of the newer sections where plantings are scattered. Sap-suckers have been known to girdle entire trees or large branches by cutting large numbers of holes in a series of lines close together around the trunk or limbs.

Morning-glory.—This is probably the worst of the weeds in almond orchards, and is the hardest to control. Sheep and chickens may be used with excellent results, provided care is taken to see that the sheep are not allowed to go hungry, for then they will bark the trunks of the trees very quickly. An excellent plan is to arrange gates so that the sheep must go through the orchard from pasture to get water. In passing back and forth they will forage over the entire orchard and dig up all the morning-glory in sight. Chickens are fond of the succulent new shoots, and will keep them below the ground until the underground stems and roots weaken and die. Cultivation throughout the growing season, often enough to prevent the morning-glory from developing any leaves for a whole year at least, and longer if necessary, will starve the plants to death.

VARIETIES

The problem of selecting varieties of almonds for planting in California is to choose the ones that are most marketable and at the same time to secure best results in cross-pollination. At the present time there are comparatively few varieties of almonds which have won and retained their popularity with the grower and the trade. There are a number of reasons for this condition. The cultural, climatic and soil conditions under which the almond thrives are much more limited than for most of the common deciduous fruits; the area of production in America is limited largely to California; the industry is comparatively new, and in general, there is not the great varietal variation in season, appearance, texture, flavor and behavior that is found in the apple, peach, pear and similar fruits. The comparatively rapid deterioration of fleshy fruits after ripening makes a succession of varieties desirable to extend the period of consumption. With almonds, the season of all varieties extends from one harvest to the next, if properly handled. Growers are, therefore, recommended to plant only standard, marketable varieties as far as possible.

Early ripening varieties must be chosen. The crop must be in the hands of the trade early, for the bulk of it is used in the holiday trade. This is especially true with unshelled almonds. Early almonds only can be harvested and sold before the bulk of the European shipments arrive.

The actual time of ripening of the different varieties is variable from season to season, and in different sections or even different orchards in the same season. Generally, harvesting commences early in August and closes about the middle of October. The approximate order of ripening of the better known varieties is shown in table IX.

TABLE IX
APPROXIMATE ORDER OF RIPENING OF VARIETIES OF ALMONDS

	1.	Nonpareil
	2.	I.X.L.
		Jordan
	3.	Ne Plus Ultra
	4.	Peerless
		Princess
		California
		King
		Silver

	5.	Golden State
	6.	Lewelling
	7.	Drake
	8.	Languedoc
		Texas

Many new varieties have been originated in California but most of them have fallen into disfavor in a short time. In fact, it is impossible to locate even single trees of some varieties which were formerly well known. From time to time, however, worthy varieties have been introduced and have succeeded in making a permanent place for themselves through their ability to fill a demand that before had been but partially or poorly supplied.

In view of the changing market situation for California almonds, due to the rapidly increasing acreage and the very limited demand for unshelled almonds, it is safe to predict that the only new varieties which will be of value in the future will be those that are primarily of superior quality for shelling purposes. Yield must take second place.

Well known varieties are not only in heavier demand in the principal markets, but they invariably bring much better prices than the newer varieties. In some years, when the domestic crop of a given variety is light and the demand good, it is possible to unload poorer or less known varieties at fairly good prices. More often, however, they are a drag on the market.

Owing to the increasing consumption of shelled almonds and the probability of a still greater increase in the future, growers should arrange future plantings with a view to supplying the best shelling varieties. At the present time, the best shelling varieties are not ordinarily the heaviest producers. With a limited production they may not even bring as large returns as the poorer but heavier yielding varieties. As the production increases, which it is doing very rapidly, the relative value of the best shelling varieties will increase in proportion and they may be sold at good prices when it will be impossible to move a heavy tonnage of a poor variety at a profitable price.

Although the future almond markets will no doubt use shelled almonds very largely, there will always be a limited demand for unshelled almonds for use in the holiday trade and for home table use. Unshelled almonds to be acceptable for such purposes must be large, attractive nuts with light-colored, clean-looking shells, soft enough to be broken with the hands. The kernels must be well filled and free from gum. The I.X.L. is the most popular and highest-priced nut for this purpose. The Ne Plus Ultra ranks next because of its attractive outside appearance and shape; one of the principal objections to it being its tendency to have gummy kernels. The Drake is another variety in demand for this purpose. It is moderately large, plump and well filled with a good quality kernel, and while not as attractive as the I.X.L. or Ne Plus Ultra, it is popular with the medium-priced trade. A certain class of trade prefers the Nonpareil for such use, and it appears to be growing in popularity because of the attractive kernel and the ease with which shelling by hand is accomplished.

The confectioners, on the other hand, care nothing for shell. They want a medium or large sized kernel, uniform in shape, and plump; one that can be coated smoothly or evenly with candy. For blanching and salting purposes, the kernels must be large and smooth. The best California variety for this purpose is the Nonpareil. It is also the best nut for table use when sold shelled. As a rule, the papershell varieties are the best for shelling because of the large percentage of unbroken kernels which may be obtained. The broken kernels and those obtained from cheaper and less desirable varieties are used largely by the bakers and almond-paste manufacturers.

The planting of large blocks of orchards to single varieties is not a wise practice. Planting of several varieties will assist greatly in lengthening the harvest season, and thus enable one to handle large crops with fewer men and less equipment. For example, the four best varieties—the Nonpareil, I.X.L., Ne Plus Ultra and Drake—ripen in the order named; the Nonpareil ripening about two weeks before the I.X.L., the Ne Plus Ultra about a week after the I.X.L., and the Drake about two weeks after the Ne Plus Ultra. Where there is danger of failure of varieties to set fruit due to frost or improper pollination or unfavorable weather conditions during, or soon after, blooming, the grower is more likely to get a crop from some variety if several are planted to secure a succession of bloom in the spring.

The principal reason for interplanting varieties is to secure adequate cross-pollination. For this purpose the Ne Plus Ultra and Drake are probably the best to use as pollenizers.

Other combinations, as indicated on page 6, may be made that will be satisfactory, though care must be exercised to secure varieties that blossom near enough together to be effective. Figure 26 shows the effective blossoming period for fifteen varieties.

Adaptation of Varieties.—The best marketable nuts are, as has been suggested, few in number, and most of these do well in all of the principal almond districts of California. Where the climatic and soil condition are equally favorable there is no great variation in their behavior, but owing to such differences it has been found that certain varieties are better adapted to some districts than others.

The Nonpareil, the best variety known at the present time for California conditions, bears more nearly uniform crops from year to year and shows a wider range of adaptation than any of the other good commercial varieties. It has proved itself to be satisfactory in every almond district in the state. The Drake closely approaches the Nonpareil in this respect. The I.X.L. and Ne Plus Ultra are the most variable in their behavior. The blossoms of the two varieties seem to be more tender and hence more liable to injury under unfavorable conditions; gumming is more prevalent near the coast, and during harvest the slower ripening and opening of the hulls in the more moist atmosphere in many of the coast valleys causes excessive darkening and sometimes molding of the shell. The Ne Plus Ultra does its best on comparatively high, well-drained soils, adjacent to the larger streams in the Sacramento Valley, such as the lands along the Sacramento River, Putah Creek, Cache Creek, etc., though it also grows and produces well in the Banning district and in many of the foothill sections where conditions are favorable. The I.X.L. does best on the foothills surrounding the Sacramento Valley, notable on the west side. In the Banning district the Ne Plus Ultra seems to be a better producer than the I.X.L. Varieties which ripen later than the Drake should be avoided in the Banning district because of the liability to damage from the frequent October rains.

There are a number of different varieties, such as the Eureka and Jordan, which give promise of filling a limited place in the markets but which have not yet been thoroughly tested throughout the state. The Eureka is popular with confectioners because of the similarity in shape to the Jordan. In limited quantities the demand is good. It is still a question as to whether it would hold up in price if grown in very large quantities. The Jordan nut is of excellent quality but in California the trees are variable in vigor. In some cases the trees make unusually large, vigorous growth while in others they are small and apparently stunted. The cause of this has never been adequately determined. In general, the Jordans do not bear sufficiently heavy crops to make them pay at the prevailing low prices. The chief reasons for the low prices are the extreme hardness of the shell and the absence of satisfactory methods of shelling. The invention of a satisfactory machine for this purpose would probably make it pay to plant Jordans in much larger quantities.

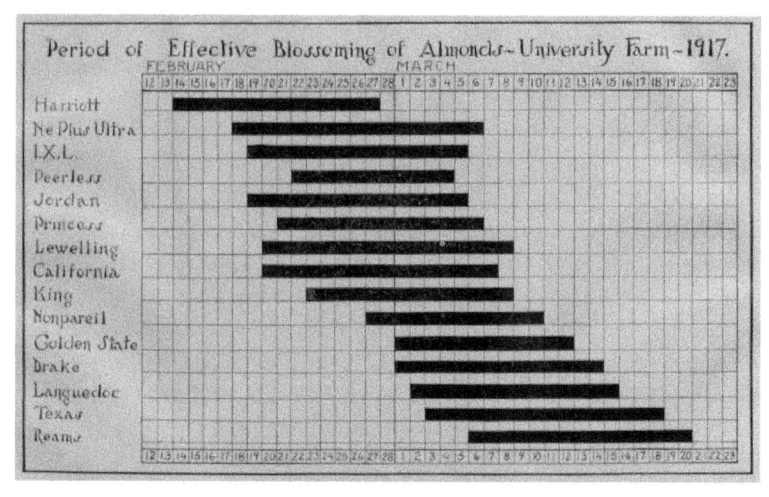

Fig. 26.—Period of Effective Blossoming of Almonds—University Farm—1917.

Other varieties, such as the Texas, have been sufficiently tested to show them to be well adapted to most districts, but they are not to be recommended because of the difficulty experienced in marketing them at a profit in large quantities. The Texas has been planted extensively in California without sufficient justification. It was planted because of its value as a pollenizer, its precocious and prolific bearing, and its upright habit of growth. In small quantities it was sold in less exacting markets as a Drake, but in larger quantities there has been a good deal of objection to it on the part of the trade. As the bearing trees become older and bear heavier crops, the nuts tend to become smaller and the shells harder, which increases the difficulty of selling. From the standpoint of the grower as well as the market, the lateness in ripening is very objectionable. The California Almond Growers' Exchange is finding it harder each year to satisfactorily market the rapidly increasing tonnage of this variety. It should, therefore, be avoided in new plantings.

SIZE OF ALMONDS

Records of fourteen varieties of almonds grown in the same orchard under similar conditions of soil and culture have been kept at the University Farm at Davis, for the years 1913 to 1916, inclusive. These trees are all of the same age, except the Texas, Peerless and Harriott, which are one year younger than the others. These figures show that there is a wide variation within varieties from year to year, not only as regards size, but proportion of shell to whole nut, and in the proportion of double kernels.

Table X shows the variation in size from year to year, as indicated by the number of nuts per pound. Five-pound samples of each variety were used to determine the average size:

TABLE X
NUMBER OF ALMONDS PER POUND FOR YEARS 1913 TO 1916, INCLUSIVE

Variety	1913	1914	1915	1916	Average
Nonpareil	256	240	224	223	236
I.X.L.	149	149	159	150	152
Ne Plus Ultra	—	163	179	169	167
Drake	189	132	131	130	145.5
Languedoc	226	220	185	213	211
Texas	182	171	165	173	173
Reams	138	123	143	148	140.5
Lewelling	154	150	158	139	150
Peerless	—	118	127	134	124
Princess	270	241	190	252	238
California	—	232	211	206	216
King	—	246	234	244	241
Harriott	—	175	151	176	167
Jordan	—	76	70	75	74

The value of an almond for shelling depends not alone on its relative ease of cracking, but also upon the percentage of kernel to the whole nut as shown in table XI:

TABLE XI
PERCENTAGE OF KERNELS TO WHOLE NUTS

Variety	1913	1914	1915	1916	Average
	Per cent	Per cent	Per cent	Per cent	Per cent
Nonpareil	67.5	65.0	67.2	67.0	66.6
I.X.L.	45.31	48.5	60.6	54.7	52.28
Ne Plus Ultra	—	53.25	57.6	58.5	56.45

Drake	46.33	42.40	47.0	42.2	44.48
Languedoc	48.75	50.0	49.4	48.5	49.16
Texas	43.75	45.0	42.4	44.4	43.89
Reams	45.94	42.5	49.25	43.4	45.27
Lewelling	43.44	48.8	50.6	45.0	46.96
Peerless	—	36.0	39.65	32.5	36.05
Princess	65.0	73.0	70.6	73.8	70.6
California	—	71.2	70.6	69.5	70.43
King	—	70.0	72.0	71.6	71.2
Harriott	—	56.25	54.2	50.6	53.68
Jordan	—	25.0	23.8	26.9	25.23

Double kernels are particularly undesirable in nuts for shelling because of the irregular shape of the halves, which renders them unfit for confectionery or bakery purposes where whole kernels are used. Table XII shows the percentage of double kernels by number:

TABLE XII
PERCENTAGE OF DOUBLE KERNELS FOR DIFFERENT VARIETIES

Variety	1914	1915	1916	Average
Nonpareil	1.09	1.96	6.10	3.05
I.X.L.	.50	.50	.80	.60
Ne Plus Ultra	4.30	5.76	12.06	7.37
Drake	12.72	6.25	6.65	8.54
Languedoc	.99	.32	1.60	.97
Texas	11.56	7.40	11.90	10.29
Reams	9.92	13.79	4.82	9.51
Lewelling	13.30	28.70	46.50	29.50
Peerless	4.44	7.05	8.38	6.62
Princess	.83	.21	4.61	1.88
California	.00	.00	.097	.032

King	.00	1.28	2.13	1.14
Harriott	1.43	.40	.80	.88
Jordan	1.32	3.72	3.74	2.93

METHODS OF CLASSIFICATION

Almonds are classified according to hardness of shell, into four classes:

Papershell.—Those almonds having a thin, papery shell which may easily be broken between the fingers of one hand.

Softshell.—Those which have a more or less spongy or thin shell which may be broken between the fingers of two hands.

Fig. 27.—Almond varieties.

LANGUEDOC. TEXAS. DRAKE.

NONPAREIL. I.X.L. NE PLUS ULTRA.

TARRAGONA. JORDAN. PEERLESS.

Standardshell.—Those requiring very strong pressure of the hand or the use of a nut-cracker to break. These may have a spongy or smooth outer shell.

Hardshell.—Those which cannot be broken by hand but require a sharp blow with a hammer or strong pressure with a nut cracker to crack them.

The papershell varieties are excellent for shelling as they contain a large percentage of kernels which may easily be obtained whole. The principal objections are that birds are particularly fond of them since they can crack them easily, and the shells are often poorly sealed. The latter fact makes it difficult to prevent worm infestation and to prevent the penetration of sulphur fumes to the kernel during the bleaching process. As a result practically none of the papershells are bleached but are sold for shelling purposes.

The softshells are generally more attractive for table use because the shells are more perfect and, in commercial varieties, are brighter in color and more attractive. The shells are usually well sealed and can be bleached to give additional brightness with less danger of the fumes penetrating to the kernel than the papershells.

Fig. 28.—Almond varieties.

HARRIOTT. EUREKA. LEWELLING.

KING. CALIFORNIA. PRINCESS.

STUART. LA PRIMA. BATHAM.

The standardshells have the greatest range in character, thickness and hardness of shell. For table use they are sometimes too hard for high-class trade. The percentage of kernel is too low to make them very attractive to the retail trade. Of this class of almond the Drake is probably the best of the

California varieties. Almond varieties of all these classes are shown in figures 27 and 28. These illustrations show most of the varieties grown to any extent in California and others that have attracted much interest and inquiry, together with the European Tarragona.

FOOTNOTES

[1] Tufts, W. P., unpublished data from experiments conducted in the University Farm orchard, at Davis, California.

[2] Colby, Geo. E., Ann. Rept. Cal. Agri. Exp. Sta., 1895-1896 and 1896-1897.

[3] As this bulletin goes to press doubt is raised as to whether some of these trees may not be on almond stock, but this in no material way affects the discussion.

[4] Just what conditions will exist after the close of the present war cannot be forecast with any degree of accuracy. One thing is certain, the increasing popularity of shelled almonds makes it essential that the American markets become familiar with the California shelled product, and that this trade be extended as quickly as possible.

[5] Cir. 121, Univ. Calif. Agri. Exp. Station, October, 1914.

[6] The fumigation of Stored Grain, Dried Fruits, and Other Products, E. R. DeOng, Cal. Agr. Exp. Sta., Dec., 1917.

BIBLIOGRAPHY

TREAT, W.

1890. Almond Culture. Report of California Board of Horticulture, 1900, pp. 72-78.

FULLER, A. S.

1896. The Nut Culturist, pp. 12-43. Orange Judd Co., New York.

CORSA, W. P.

1896. Nut Culture in the United States, Embracing Native and Introduced Species. U. S. D. A., Division of Pomology, pp. 19-28.

COLBY, GEO. E.

1898. Analysis of California Almonds. University of California Agr. Exp. Sta. Report 1895-1896; 1896-1897, pp. 145-151.

FAIRCHILD, DAVID G.

1902. Spanish Almonds and Their Introduction into America. Bulletin 26, Bureau of Plant Industry, U. S. D. A., pp. 7-14 and 8 plates.

DARGITZ, J. P.

1909. The Almond Commercially Considered. Proc. 36th Cal. State Fruit Growers' Convention, pp. 64-71.

WICKSON, E. J.

1914. California Fruits and How to Grow Them. 7th edition., pp. 424-430.

BAILEY, L. H., and WICKSON, E. J.

1914. Almond, Standard Cyclopedia of Horticulture, vol. I, pp. 249-251.

HUNT, THOS. F., and Staff.

1914. Some Things the Prospective Settler Should Know. Circular 121, Cal. Agr. Exp. Station, pp. 3, 8, 41-42.

TAYLOR, R. H.

1915. Present Status of the Nut Industry in California. Proceedings of the Society for Horticultural Science, 1915, pp. 31-39.

1915. A Symposium of California Pomology: The Almond. Proceedings American Pomological Society, 1915, pp. 121-126.

PIERCE, GEO. W.

1915. The Status of the Almond Industry of the Pacific Coast. Proceedings American Pomological Society, 1915, pp. 75-82.

STEUBENRAUCH, A. V., and TAYLOR, R. H.

1915. Some Lessons from the California Nut Industry. Proceedings 14th Ann. Conv. National Nut Growers' Association, 1915, pp. 90-93.

www.ingramcontent.com/pod-product-compliance
Ingram Content Group UK Ltd.
Pitfield, Milton Keynes, MK11 3LW, UK
UKHW031338260325
456749UK00002B/331